The Conceptions of Reality in the Interpretations of Quantum Mechanics

Gerd Christian Krizek

The Conceptions of Reality in the Interpretations of Quantum Mechanics

A Journey from Aristotle's Principles to Einstein's Dreams of a Unified Theory

Springer

Gerd Christian Krizek
Faculty of Physics
University of Vienna
Vienna, Austria

ISBN 978-3-031-65785-6 ISBN 978-3-031-65786-3 (eBook)
https://doi.org/10.1007/978-3-031-65786-3

This Springer imprint is published by the registered company Springer Nature Switzerland AG
The registered company address is: Gewerbestrasse 11, 6330 Cham, Switzerland

Preface

The first seeds for this book were sown in the Ph.D. thesis handed in by the author at the Faculty of Physics of the University of Vienna in 2017. Since then, the foundations of quantum mechanics received further attention. Many papers with new insights have been published, and Nobel prizes have been awarded, demonstrating once again the relevance of foundations of quantum mechanics. In parallel, the impact of quantum information theory in applications is evolving all the time, with new technologies to implement physical qubits, new quantum computing architectures, and new algorithms, while quantum cryptography and quantum key distribution are moving forward in leaps and bounds. At the same time, these applications of quantum information theory should not blind us to the fact that there are still a lot of open foundational questions. Some of these questions go back to the very first debates in quantum mechanics, some refer to philosophy of physics, and others refer to general philosophical questions. The aim of this book is to take the reader on a journey from Aristotle, to medieval philosophy, to Einstein's attempts to unify physics. It will be clear that the quest for answers in the foundations of quantum mechanics is not over. We may still have much work to do to achieve an understanding of the nature of reality, since quantum mechanics continues to give us new insights that we do not seem to have fully grasped yet.

The order of the chapters has been altered to provide a better reading experience, and several parts of the original thesis have been updated. Some aspects have been published in recent papers Krizek and Mairhofer (2023), Santo and Krizek (2023). Note that the thesis itself was written as a cumulative structure of several papers, which means inevitably that there will be a certain degree of repetitions between chapters.

I would like to express my sincere appreciation to Beatrix Hiesmayr, Franz Embacher, Agnes Rettelbach, Isabella Wöber, Karin Langer, and Flavio Del Santo for their important contributions to this manuscript in supervisions, discussions, graphics design, and proofreading. Your expertise, dedication, and collaborative efforts have significantly enhanced the content and depth of the book.

I would also like to thank Marian Kupczynski, Chris Fuchs, and Andrei Khrennikov for their invaluable input and discussions.

Vienna, Austria Gerd Christian Krizek

Contents

1 Introduction 1

2 A Classification Scheme for Interpretations of Quantum Mechanics 5
- 2.1 The Classification Scheme and Its Application 5
- 2.2 Application to Physics and the Interpretations of Quantum Mechanics 7
  - 2.2.1 Classical Particle Mechanics 8
  - 2.2.2 De Broglie-Bohm Interpretation 11
  - 2.2.3 Bohmian Mechanics 16

3 Philosophy of Science from Artistotle and William of Ockham to Structuralism 21
- 3.1 William of Ockham's Philosophy in Historical Context 21
  - 3.1.1 The Omnipotence Principle 22
  - 3.1.2 The Contradiction Principle 22
  - 3.1.3 The Parsimony Principle 28
- 3.2 Ockham's Razor and Other Heuristics 31
  - 3.2.1 Mach's Economy of Thought 31
  - 3.2.2 The No Alternative Argument 32
  - 3.2.3 The Meta Inductive Argument 33
  - 3.2.4 The Unexpected Explanation Argument 33
  - 3.2.5 The Ontological Coherence Argument 34
- 3.3 The Structure of Scientific Theories in the Framework of Philosophy of Science 35
  - 3.3.1 Einstein's Model of Layers 35
  - 3.3.2 The Received View 37
  - 3.3.3 The Semantic View 39
  - 3.3.4 The Structuralist Approaches 40
  - 3.3.5 Epistemological and Ontic Structural Realism 45
  - 3.3.6 Quine-Duhem Thesis 48

4 The EPR Debate, the ER Connection, and Completeness 51
4.1 Einstein's View on Theories and Theory Formation 51
4.2 Einstein's Not-Naive Realism and the Notions of Reality 54
4.2.1 Naive Realism 55
4.2.2 Subtleties in the EPR Paper 57
4.2.3 Notions of Reality 58
4.3 EPR=ER in 1935 and Its Revival 69
4.3.1 Nonlocality 69
4.3.2 Localizability of Particles 70
4.3.3 Completeness 71
4.3.4 The Unified Field Theory Program 79
4.3.5 Recent Claims of EPR = ER 82
4.4 The EPR Paper Revisited 82

5 The Concept of Reality in the Interpretations of Quantum Mechanics 85
5.1 Reality in Classical Physics 85
5.1.1 The Structure of a Physical Theory and Its Metaphysics 86
5.1.2 Uniqueness 87
5.1.3 Independence 88
5.1.4 Completeness and Correspondence 89
5.2 Reality in Contemporary Philosophy of Science 90
5.2.1 Metaphysical Realism 90
5.2.2 Scientific Realism 92
5.2.3 Object Conception and Ontology 94
5.2.4 Structural Realism 96
5.3 Concepts of Reality in Quantum Mechanics 97
5.3.1 Naive Realism 97
5.3.2 Einsteins Reality Criterion 98
5.3.3 Ruarks Reality Criterion 99
5.3.4 Local Realism in the Bell Inequalities 99
5.3.5 Reality in the Measurement Process 101
5.3.6 Structural Realism in Quantum Mechanics 102
5.4 Ockham's Razor and the Interpretations of Quantum Mechanics 110
5.4.1 Overview of Interpretations Used in the Discussion 110
5.4.2 In Favor of de Broglie-Bohm Interpretation and Bohmian Mechanics—Against Copenhagen Interpretation 112
5.4.3 In Favor of Copenhagen Interpretation—Against de Broglie-Bohm Interpretation and Bohmian Mechanics 113
5.4.4 In Favor of Copenhagen Interpretation—Against Many Worlds Interpretation 114

5.4.5 In Favor of Many Worlds Interpretation—Against de Broglie-Bohm Interpretation and Bohmian Mechanics 114
5.4.6 In Favor of de Broglie-Bohm Interpretation and Bohmian Mechanics—Against Many Worlds Interpretation 114
5.4.7 Other Positions 115
5.4.8 Analysis of Applications of Ockham's Razor 116
5.4.9 Ockham's Razor and Economy of Thought 117
5.4.10 A New Diversity 118
5.5 Statements About the Ontological Level 120
5.5.1 Ontological Coherence 120
5.5.2 Ontological Truth 123

6 Summary, Conclusion and Outlook 127

Appendix 141

References 157

Chapter 1
Introduction

One of the successes of modern physics is the methodology it uses to develop quantitative theories from phenomena. These theories combine predictive and explanatory power. Its predictive power enables technological progress, but also provides a justification for the scientific worldview given by a specific picture of reality. In this way, our best and most successful theories provide the framework for the dominant worldview of physics and philosophy. Indeed, philosophy and physics are deeply related to each other. Physics is a methodologically constrained continuation of philosophy of nature, a specific branch in the broader enterprise of science. Philosophy advocates for unrestricted freedom in methodological choices, while physics is methodologically defined by philosophy of science. The progress in philosophy and the progress in physics are strongly entangled; both fields inspire each other and move forward together.

The conceptions of reality that emerge from classical physics are dominated by common-sense realism. Space, time, and separability of objects provide the framework for describing phenomena, and this classical description of phenomena is furthermore dominated by materialistic ideas.

But with the advent of quantum mechanics, a new theory developed with an interpretation that was incompatible with the conceptions based on classical physics, while the latter did not per se show a deep difference to our daily experience. The world-view implied by this new theory was vastly different from the classical one. Since the development of quantum mechanics in the 1920s coincided with the zenith of logical empiricism in philosophy, the latter had a considerable influence on the conceptions of reality that were adopted in it.

The motivation for Einstein, Bohr, Bell and others involved in the debates on the foundations of quantum mechanics was to search for an appropriate interpretation of its mathematical formalism. Notwithstanding the empirical success of the theory, its interpretation, implications, and ontology offered several possibilities of interpretation. It has been commonly accepted as a methodology of metaphysics, that the best of our scientific theories provide insights on the nature of reality. By identifying conceptions or elements in the theory, which refer to ontological entities, the

G. C. Krizek, *The Conceptions of Reality in the Interpretations of Quantum Mechanics*,
https://doi.org/10.1007/978-3-031-65786-3_1

methodology was highly successful in classical physics. With the rise of quantum mechanics this idea was suddenly challenged.

Einstein and Bohr had a series of public disputes on the interpretation of quantum mechanics about realism and positivism, which became famous as the Bohr-Einstein debates. Although these debates influenced the development of quantum mechanics substantially, they provided no coherent ontology, and more questions than answers arose. As a result of this unsatisfactory progress, the discussion of the foundations of quantum mechanics and the nature of reality was put on hold for a time.

As quantum mechanics celebrated technological successes in the middle of the twentieth century, interest in the foundations of quantum mechanics resurfaced, with re-examinations of the Einstein-Podolsky-Rosen (EPR) paradox, and many new interpretations of quantum mechanics. Bell's inequality, Kochen-Specker theorem, and other formal theorems were found to treat foundational questions. All of them challenged the traditional conceptions of reality.

With the advent of quantum mechanics, the conceptions of reality became the subject of intense disputes and, the situation has still not been resolved. On the contrary, many different interpretations of quantum mechanics have developed over the years, involving a variety of metaphysical positions and this reflects the extent and diversity of the debate about the foundations of the theory. It turned out that The variety of possible interpretations of quantum mechanics, overlap in their empirical predictions, but are largely orthogonal in their ontological commitments.

Most of these interpretations will be presented or at least mentioned in this work. They share the same mathematical formalism, but differ in their conceptual approach and their ontology. Some interpretations contain additional elements or concepts, or they are reconstructions of quantum mechanics. It is sometimes a delicate matter to distinguish an interpretation of quantum mechanics from an extension or continuation of the theory. Strictly, one only refers to an interpretation if the mathematical formalism of two versions overlaps significantly or is identical. Most interpretations introduce additional conceptions regarding various aspects of the theory; this can involve extensions of the formalism, such as in Bohmian mechanics, or conceptual differences, such as in the Many Worlds interpretation, or different positions regarding the status of certain mathematical entities, such as the wave-function.

Reconstructions of quantum mechanics are to be distinguished from interpretations because they are, in principle, new theories; but they reproduce the predictions, formalism, and dynamics of quantum mechanics, but starting from modified foundational principles. However, since an interpretation and their specific conception can be the basis for reconstructions, the latter are closely related to interpretations of quantum mechanics and, therefore, are not sharply determined.

Since the field of interpretations of quantum mechanics remains ambiguous about the nature of reality for around ninety years now, leaving us with a variety of competing conceptions, the answer must at least be very hard to find. And indeed, in the present work, the intention is not to provide a full and conclusive answer to this question. The connecting research theme in this book is an examination of the way the term "reality" is defined in the different interpretations and their related philosophical programs. In effect, we ask what can be said about the nature of reality by studying the different interpretations of quantum mechanics and their interrelations. From this variety of interpretations of quantum mechanics, their interrelations, and their embedding in different philosophical frameworks, we can thus expect to learn more about the nature of reality.

We begin the discussion by an analysis of the difficult relation of various interpretations of quantum mechanics to each other, since one aspect of the difficulties in the interpretations of quantum mechanics is the mutual critique between different interpretations. One particular conception that is often involved in these discussions is the concept of Ockham's razor. We provide an overview of the historical origin of Ockham's razor and its application in quantum mechanics, and ask whether Ockham's razor is a suitable tool decide between one interpretation and another.

In order to give a proper account on the different interpretations of quantum mechanics, we propose a classification scheme and associated methodology for speaking about mathematical, interpretational, conceptional, and ontological aspects of those interpretations of quantum mechanics. We analyse the classification scheme in light of the debate on the structure of scientific theories that continued through the twentieth century, from logical empiricism to structural realism. We illustrate the classification scheme by means of classical particle mechanics, and apply it to the field of interpretations of quantum mechanics using the example of the de Broglie-Bohm interpretation and Bohmian mechanics. The classification scheme demonstrates the differences in the conception, the mathematical formalism and the ontology of these two interpretations of quantum mechanics.

Concerning the ontological level of the classification scheme, we propose two conceptions that are helpful in context with ontological statements of physical theories, namely the Ontological Coherence Argument and the concept of ontological truth. The Ontological Coherence Argument is a heuristic that helps us to decide which theories should be adopted an the basis of their ontological relations to each other. Concerning the truth values of scientific theories, we propose a theorem on ontological truth, which states that scientific knowledge is something cumulative, but not convergent.

Using the structure of this classification scheme we analyse Einstein's view on realism and the motivation for the EPR argument. In this context, we define the different conceptions of completeness utilized in the foundations of quantum mechanics. Several conceptions of realism are also involved in the discussion of the EPR argument and Einstein's ideas of reality, and this leads to an analysis of the conceptions of realism in classical physics, the philosophy of science in general, and quantum mechanics.

We provide an overview of the conceptions of reality involved in the different interpretations and their connection with recent conceptions in the philosophy of science. Structural realism is presented as a philosophical concept which has received a lot of attention in the context of quantum mechanics in the past twenty years, and which attempts to provide a coherent position on the question of realism in quantum mechanics. Since not all interpretations of quantum mechanics have been expressed in the context to structural realism, we amend the existing presentation with three interpretations of quantum mechanics that emphasize information as a foundational conception, but each in its own way.

As we conclude this introduction, our journey through the diverse landscape of quantum mechanics will not only enrich our understanding of the nature of reality and its limits but also illuminate the intertwined relationship between philosophy and physics. This exploration into the various interpretations of quantum mechanics, each with its distinct ontological commitments and conceptual frameworks, highlights the profound richness and diversity in our quest to understand the fabric of reality. While this work does not claim to resolve the deep questions that quantum mechanics poses about the nature of reality, it serves as a testament to the intellectual vibrancy and the enduring curiosity that drive both physics and philosophy and will provide valuable insights. Now, let us embark on a journey that not only reflects on the theoretical achievements of quantum mechanics but also on its philosophical relevance, inviting us to ponder the nature of reality through a multidisciplinary lens.

Chapter 2
A Classification Scheme for Interpretations of Quantum Mechanics

Quantum mechanics is undoubtedly one of the most successful theories in physics, but its foundations and interpretations are still a heavily disputed topic. Quantum mechanics allows for various interpretations, but it is surprising how significant the differences can be concerning the ontological commitments of these interpretations. Before we can discuss these differences in the interpretations of quantum mechanics we propose a classification scheme that will provide a methodology to analyse and distinguish the differences of several interpretational approaches in the mathematical model, the interpretation of the respective mathematical model, the used concepts, and the ontology. The classification scheme provides a number of levels of a theory or interpretation to distinguish between these aspects.

The scheme does not assert to represent any ontological aspects on the nature of theories itself. Although the levels and boundaries could be defined differently, this proposed classification scheme is valuable for contrasting the content of various approaches in interpreting quantum mechanics.

We present the relation between the classification scheme and approaches to the structure of scientific theories in philosophy of science. A physical theory also rests on preconceptions and premises. To illustrate this, a brief overview of philosophical stances, in context with the conception and structure of physical theories, is presented.

Connected to the ontological level of the classification scheme, two arguments are proposed that refer to the ontologies of theories and their interrelation.

We demonstrate the merit of the scheme by means of an application on classical particle mechanics and on different interpretations of quantum mechanics.

2.1 The Classification Scheme and Its Application

Interpretational debates as part of the scientific method are as old as science itself. But no interpretational debate in physics attracted more attention than the debate about interpretations of quantum mechanics (Landsman 2006). The reason is probably that

G. C. Krizek, *The Conceptions of Reality in the Interpretations of Quantum Mechanics*,
https://doi.org/10.1007/978-3-031-65786-3_2

it is more than just a debate on different models about a physical situation. It is a debate with pervasive philosophical implications.

The field of interpretations of quantum mechanics seems to be growing continually, without reconsideration. A quote by David Mermin reflects this situation (Mermin 2012):

> Quantum mechanics is the most useful and powerful theory physicists have ever devised. Yet today, nearly 90 years after its formulation, disagreement about the meaning of the theory is stronger than ever. New interpretations appear every year. None ever disappear.

To provide a better overview on the interpretations of quantum mechanics, we propose a classification scheme that categorizes the structure of a physical theory, particular an interpretation of quantum mechanics. A physical theory consists roughly of the mathematical formalism and an interpretation, which contains the definition of symbols, theoretical terms, measurement assignments, concepts and principles, and an ontology. We refine this coarse breakdown into a four-level classification.

Definition 2.1 Level 1 of a physical theory consists of a set of mathematical quantities. These mathematical quantities are related to each other by mathematical formalism.

Definition 2.2 Level 2 of a physical theory relates a set of measurement mappings to the mathematical quantities of Level 1. It describes how the defined quantities in the mathematical description of the theory are related to sensations and experience. It relates mathematical objects to theoretical terms and experimental practice.

Definition 2.3 Level 3 of a physical theory contains a number of concepts and principles that are related to the mathematical description of the theory. These principles can be formulated as statements or logical expressions.

Definition 2.4 Level 4 of a physical theory involves the ontology of the theory and the metaphysical entities or concepts that are introduced by the theory.

We make use of the following figurative elements in Table 2.1 to visualize the classification scheme.

Through the representation of dots, we illustrate a physical quantity, whether it relates to the characteristics of a physical entity or any other measurable aspect introduced by the theory. On Level 1 of the classification scheme, a physical quantity is basically a mathematical symbol. The dots are interconnected by mathematical relationships, embodying the laws governing the interplay of physical quantities.

On the second level, the physical quantities receive their names and relation to empirical results, the measurement mappings. By that, the pure mathematical formalism becomes a physical law. The physical quantities and the laws involve concepts that are set up by theoretical terms and governed by principles. These are depicted by the areas spanned by some physical quantities and laws, and belong to Level 3 of the classification scheme.

Table 2.1 Figurative elements used in the classification scheme

	Physical quantity
	Law between physical quantities
	Concept connecting laws and physical quantities
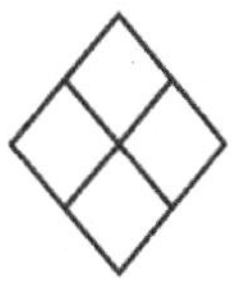	Elements of objective reality

On the ontological level, Level 4, there are entities that are a priori unknown. A physical theory may provide implicit or explicit statements on the ontology, statements that provide an insight on the metaphysical reality behind the empirical content. These statements usually involve entities that are claimed to exist independently of our perception; we denote them as elements of objective reality and depict them by a diamond.

Applying these figurative elements to the classification scheme, we can give an overview on the scheme in Table 2.2:

2.2 Application to Physics and the Interpretations of Quantum Mechanics

In this section we apply the classification scheme to several cases including some interpretations of quantum mechanics to demonstrate its practical relevance. This section can be skipped in case the reader wants to proceed to the core contents.

Table 2.2 Classification scheme: We distinguish four levels of a theory to provide a categorization to analyze different notions of completeness

1	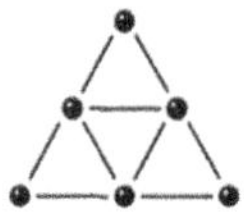	Mathematical laws between mathematical symbols
2	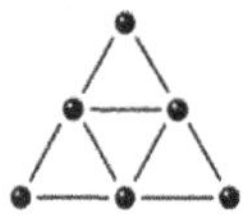	Interpretation physical quantities Relation physical quantities - Measurements Measurement laws
3	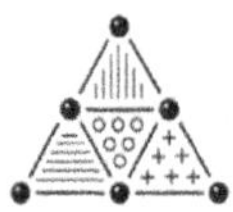	Concepts and principles
4		Ontology - beables - elements of reality

2.2.1 *Classical Particle Mechanics*

If we want to apply the classification scheme to specific physical theories and the interpretations of quantum mechanics, it is useful to demonstrate the principles by means of a well-known physical theory. We have chosen classical particle mechanics.

A detailed presentation of classical particle mechanics by means of Sneed's structuralist approach has been provided by Ishigaki (1995). We demonstrate the application of the classification scheme in a brief, incomplete, but illustrative way.

Level 1—The Mathematical Formalism

The mathematical formalism is set up by vector valued functions $\vec{x}$ and $\vec{F}$ in the vector space $\mathbb{R}^3$ and a scalar m. The vectors or rather their components are in generality functions of the free parameter time. The vector $\vec{x}(t)$ has a derivative $\frac{d\vec{x}}{dt}$ we denote with $\vec{v}$.

$$\frac{d(m\vec{v})}{dt} = \vec{F} \tag{2.1}$$

This is a linear second order differential equation. A solution is given by $\vec{x}(t)$.

Level 2—Physical Quantities and Connection to Empirical Rules

In the Level 2 of the classification scheme, the physical quantities and their representation by the mathematical symbols introduced in Level 1 are defined.

Definition 2.5 The physical quantity "position" is represented by the vector $\vec{x}$.

Definition 2.6 The physical quantity "time" is represented by the parameter t.

Definition 2.7 The physical quantity "mass" is represented by the scalar m.

Definition 2.8 The physical quantity "force" is represented by the vector $\vec{F}$.

The physical quantity "position" represents results of measurements of the position coordinate of an object; for an extensive object, this will be the center of mass coordinates, in case of a pointlike object, it will be the coordinates of the point.

We adopt the SI-unit system to relate these physical quantities with empirical statements.

One peculiarity is that we cannot define a measurement procedure for the physical quantity force independently of the acceleration of the object unless we introduced a source of a force, such as Newton's gravitational law, which we will omit.

Level 3—Concepts and Principles

We define the concept of a "point particle" to describe the behaviour of extensive objects and pointlike entities as well:

Definition 2.9 A point particle is defined by its position coordinate in space represented by a vector $\vec{x}$ in the vector space $\mathbb{R}^3$, the parameter time t as well as a set of associated properties. In the case of classical particle mechanics, this set is containing the physical quantity "mass."

The full power of the formalism of classical particle mechanics is available when we relate it with other specific laws of force. Therefore, the conception of a force is needed.

Definition 2.10 A force is an interaction between an object and the origin of the specific force. The object will change the motion of the object according to the defined law, depending on the quantity of the force and its set of properties.

With that definition of force, concrete laws of force can be introduced, such as the gravitational law, Hooke's law, or friction. This interrelates the theory to other theories, which is the idea of the conception of links in Sneed's structuralist approach.

Level 4—Ontology

The ontological statements on classical particle mechanics involve connected ontologies of space and time and involves them in the ontological level of classical particle mechanics.

Classical particle mechanics is based on a particle and matter ontology in which there are entities named particles, made of matter, that exist in a pre-existing space and alter their positions due to a time parameter according to influences and interactions by forces. The forces are on the same ontological footing, in that sense as real, as the particles and represent the interaction process between particles with other entities. This ontology is deeply related to materialism, which postulates that what exist in the ontology, is matter in motion in a pre-existing space and time framework.

Overview

See Table 2.3.

Table 2.3 Classical particle mechanics

1	$m\ddot{x} = \sum_{i=1}^{N} \vec{F}_i$
2	x... center-of-mass coordinate of an object m ... intertial mass of an object $\sum_{i=1}^{N} \vec{F}_i$ ... sum of all forces Measurement rules for $\vec{F}, m, \vec{x}(t), t$
3	Point particle concept
4	Entity of matter—Materialism Position in space and time

2.2.2 De Broglie-Bohm Interpretation

The De Broglie-Bohm interpretation is a realistic, deterministic, nonlocal[1] hidden-variable interpretation and is based on the decomposition of the Schrödinger equation into a set of two equations. One equation is interpreted as the conservation of probability, and the other equation is interpreted as a Hamilton Jacobi equation with an extra term. This term is interpreted as a Quantum Potential. From the Hamilton Jacobi equation, an equation of motion for particles can be derived, which is called guidance equation. In the De-Broglie Bohm interpretation, nonlocality is an important and inherent feature of quantum mechanics, but is only an aspect of a much deeper principle that is central to the De Broglie-Bohm interpretation, the concept of ontological holism. The De Broglie-Bohm interpretation must be distinguished from Bohmian mechanics, which is based on the same formalism, but adheres to a different conception and ontology. We present Bohmian mechanics in Sect. 2.2.3. For a detailed presentation of the De Broglie-Bohm interpretation, refer to Bohm and Curd (1981), Holland (1995).

Level 1—The Mathematical Formalism

$$\psi : \mathbb{R}^{3N} \times \mathbb{R} \to \mathbb{C} \tag{2.2}$$

$$\psi = \psi(\vec{q}_k, t) \tag{2.3}$$

$$k = 1, ..., N \tag{2.4}$$

$$\psi = Re^{i\frac{S}{\hbar}} \tag{2.5}$$

$$R : \mathbb{R}^{3N} \times \mathbb{R} \to \mathbb{R} \tag{2.6}$$

$$S : \mathbb{R}^{3N} \times \mathbb{R} \to \mathbb{R} \tag{2.7}$$

$$i\hbar\frac{\partial\psi}{\partial t}(\vec{q}, t) = -\sum_{k=1}^{N} \frac{\hbar^2}{2m_k}\Delta_k\psi(\vec{q}, t) + V(\vec{q})\psi(\vec{q}, t) \tag{2.8}$$

$$\frac{\partial S}{\partial t} + \sum_{k=1}^{N} \frac{(\nabla_k S)^2}{2m_k} + \sum_{k=1}^{N} -\frac{\hbar^2\nabla_k^2 R}{2m_k R} + V = 0 \tag{2.9}$$

$$\frac{\partial(R^2)}{\partial t} + \sum_{k=1}^{N} \nabla_k \cdot \frac{R^2\nabla_k S}{m_k} = 0 \tag{2.10}$$

[1] Locality is the concept that events are independent of all other events that are space-like separated. Nonlocality abolishes this restriction, and therefore events can have an instantaneous influence on space-like separated events, but superluminal signalling is still restricted in this interpretation.

Level 2—Physical Quantities and Connection to Empirical Rules

The mathematical level contains the Schrödinger equation. Every complex valued function can be written in a polar form of the kind in Eq. (2.6) with two real valued functions, R and S. Applying the polar form on the Schrödinger equation, one obtains Eqs. (2.9) and (2.10). This is just a reformulation of the Schrödinger equation.

Interpreting Eq. (2.9), one sees that it resembles the Hamilton-Jacobi equation of classical mechanics. A detailed account of Hamilton-Jacobi theory in the context of the De Broglie-Bohm interpretation is provided by Holland (1995, p. 27). The term S resembles the action of classical mechanics, but there is an additional term that has no counterpart in classical Hamilton-Jacobi mechanics; it is the term we denote by Q and is usually named Quantum potential:

$$Q = \sum_{k=1}^{N} -\frac{\hbar^2 \nabla_k^2 R}{2m_k R} \tag{2.11}$$

The Quantum potential is interpreted on the same footing as a potential energy, but due to its structure, it has some specific properties that are emphasized as in-formational (Holland 1995, p. 89):

> *Thus contary to what one might expect in a classical wave, a particle does not respond to the intensity of the wave in its vicinity, but rather to its form.*

The proponents of the De Broglie-Bohm interpretation often make use of a radio-wave analogy for the Quantum potential and the in-formational character. The radio waves that control the movement of a remote-controlled vehicle carry an informational content that is independent of the amplitude of the radio wave. Essential is its informational content that guides the remote controlled object. In the sense needed on Level 2 we interpret the term Q as a potential, equally important to the guidance of the particle as the classical potential §V.

By interpreting Eq. (2.9) as a Hamilton-Jacobi type equation in the De Broglie Bohm interpretation, one concludes that there are equations of motions for the individual systems (Holland 1995):

> *... let us construct a vector field ... and assume that the latter defines at each point of space at each instant the tangent to a possible particle trajectory passing through that point. This naturally provides a description of an ensemble of particles, which is fictitious in the sense that only one track is realized in any given field ...*

For one particle, this vector field is given by

$$\frac{d\vec{q}}{dt} = \frac{\vec{p}}{m} = \frac{\nabla S}{m}. \tag{2.12}$$

This equation is called guidance equation and provides a set of trajectories as solutions. In a many-particle situation, the guidance equation is given by

$$\frac{d\vec{q}_k}{dt} = \frac{\nabla_k S}{m_k}. \tag{2.13}$$

The $\vec{q}_k(t)$ represent the positions of the particles at specific times t.

The interpretation of the function R is consistent with the Born interpretation in some aspects, but not in all (Bohm et al. 1987, p. 325): In that sense Eq. (2.10) ...

> *... can evidently be regarded as a continuity equation with $P = R^2$ being a probability density, as Born suggested. The function P has, however, two interpretations, one through the Quantum Potential and the other through the probability density. It is our proposal that the fundamental meaning of R (and therefore indirectly of P) is that it determines the Quantum Potential. A secondary meaning is that it gives the probability density for the particle to be at a certain position. Here we differ from Born who supposed that it was the probability of finding the particle there in a suitable measurement.*

Level 3—Concepts and Principles

Particle and wave conception: A particle is described in the $\mathbb{R}^3$ and is defined by its particle trajectory $q_k(t)$. A particle is guided by the wave function or two real valued components R and S (Holland 1995, p. 277):

> *When we speak of a "many body system" we mean then a single wavefunction together with a set of particles. There is in general no wave associated just with each particle individually.*

It is important to emphasize this aspect of the many particle situation in the De Broglie-Bohm interpretation, as an essential point gets lost if one studies only the single particle situation. In the single particle case, the configuration space is a $\mathbb{R}^3$ and isomorphic to the physical space $\mathbb{R}^3$. This gives the impression that the De Broglie-Bohm interpretation is acting on the physical space. In fact, it does, but only the positions $q_k(t)$ refer to physical space, and the functions R and S and therefore the ψ live on the $\mathbb{R}^{3N}$ configuration space.

In the De Broglie-Bohm interpretation, the wave function has no primary conceptual status; it is decomposed in the two real valued functions R and S. Nevertheless, the De Broglie-Bohm interpretation refers also to the wave function and outlines characteristics. We follow the presentation by Holland (1995, p. 83–84) and present those aspects relevant to our discussion:

- There is no source of the ψ-field. It has no localized origin.
- There is no ontological construction for a medium where the ψ-function oscillates on.
- Due to the formal construction of the Quantum potential influences by the ψ-function are independent of the intensity and therefore nonlocalities are easily realised.
- Due to the formal construction of the Quantum potential, influences by the ψ-function are independent of the intensity of the ψ-function respectively of the function R and therefore nonlocalities are easily realised.

Hidden variables: The De Broglie-Bohm interpretation is a hidden variable theory; these hidden variables are the particles positions $\vec{q}_k(t)$ and their initial values $\vec{q}_o$.

Quantum potential: One conception is emphasized in the De Broglie-Bohm interpretation. It is the conception of the Quantum potential represented by Eq. (2.11). The Quantum potential plays the key role in the conceptual apparatus of the De Broglie-Bohm interpretation (Holland 1995, p. 78):

> *The introduction of the quantum potential as a causal agent has explanatory power which one unnecessarily forgoes by concentrating on just (3.2.19).*[2] *It represents the difference between classical and quantum mechanics.*

Though proponents of the De Broglie-Bohm interpretation claim a resemblance of the mathematical apparatus with classical mechanics, particularly Hamilton-Jacobi mechanics, they do not aim for a return to classical conceptions or ontology. The Hamilton-Jacobi formalism is merely a unified language that speaks about phenomena in classical and quantum context (Holland 1995, p. 78):

> *The radical departure from classical notions inherent in the causal interpretation*[3] *lies in the details of the model, in the types of motions that are accessible to particles.*

Level 4—Ontology

Particle and wave ontology: The particle ontology of the De Broglie-Bohm interpretation leads to a new conception of matter (Holland 1995, p. 65):

> *The classical theory of matter, ... has been replaced by a more general conception in which matter has an intrinsic field aspect, the mass points moving and interacting under the influence of the new internal energy as well as the more familiar potentials of classical dynamics. This leads to a synthesis of the wave and particle characteristics of matter. These properties are not mutually exclusive, but simultaneously existent.*

A particle is not only an entity represented by a set of properties and a position. In the De Broglie-Bohm interpretation, the particle conception goes beyond that classical naive representation of particles (Bohm et al. 1987):

> *As the theory develops, we shall find that the electron is by no means a structureless particle. Rather, what is suggested by its behaviour is that it is a highly complex entity that is deeply affected by its quantum field in an extremely subtle and dynamic way. Moreover, this entity is not to be regarded (as is done in the usual interpretations) as somehow directly possessing both particle-like and wave-like properties. Rather, the observed wave-like properties will follow, as we shall see, from the general effect of the quantum wave field on the complex structure of the particle.*

The particle wave ontology is not a simple replacement of the classical wave and particle pictures by an assertion that quantum systems behave in some situations wave-like and in some situations particle-like. It is more; it is the view that matter

---

[2] The guidance equation.

[3] The De Broglie-Bohm interpretation is often referred to as the causal interpretation.

behaves in a novel way in the quantum regime that is incompatible with both ontological pictures of particle and waves alike. On the question, what are the quantum objects, an appropriate answer would be: something different.

Active information and the causa formalis: The De Broglie-Bohm interpretation emphasizes a conception of in-formation that is related to the Aristotelian causa formalis (Bohm and Curd 1981, p. 16):

> *In more modern language, it would be better to describe this as formative cause, to emphasize that what is involved is not a mere form imposed from without, but rather an ordered and structured inner movement that is essential to what things are*

By proponents of the De Broglie-Bohm interpretation this term was used to describe this feature in sense of an in-formation: Active information. Hiley and Pylkkanen (2005, p. 19) describes its features:

> *The proposal is that active information at the quantum level organizes the dynamical evolution of the system itself. ... As we remarked earlier, the quantum potential appears to be some kind of internal energy which carries information about the environment. Therefore we consider the idea that the whole process, particle plus active environment (which in terms of a measurement requires the specification of experimental conditions), is being formed partly from within. This suggests that the process is more organic than mechanical.*

Active information can be understood in the sense of the radio-wave remote-controlled vehicle. The vehicle's motions depend not only on its own local laws, but on some information on the environment provided by a formative cause.

A recent account on the role of the Quantum potential in the context with Active information is provided by Dennis et al. (2015).

Ontological Holism: The De Broglie-Bohm interpretation advances an ontological holism. The Quantum potential reflects this holism and is the element in the theory that acts in the sense of this wholeness (Bohm et al. 1987, p. 340):

> *We have seen thus far that the quantum behaviour of matter shows a certain kind of wholeness, brought about by the quantum potential. This latter functions as active information that may reflect distant features of the environment and may give rise to a non-local connection between particles that depends on the quantum state of the whole, in a way that is not expressible in terms of the relationships of the particles alone.*

According to the ontology advanced by the De Broglie-Bohm interpretation, there is a wholeness in the very nature of reality that is hidden to us in some branches of phenomena, like classical physics, but revealed by Quantum theory. Wholeness in that respect affects also the weight and relevance of our scientific theories (Bohm and Curd 1981, p. 21–p. 22):

> *We thus to be alert to give careful attention and serious consideration to the fact that our theories are not descriptions of reality as it is, but rather ever-changing forms of insight, which can point to or indicate a reality that is implicit and not describable or specifiable in its totality.*

Table 2.4 Classification scheme of the De Broglie-Bohm interpretation

1	$\frac{\partial S}{\partial t} + \sum_{k=1}^{N} \frac{(\nabla_k S)^2}{2m_k} + \sum_{k=1}^{N} - \frac{\hbar^2 \nabla_k^2 R}{2m_k R} + V = 0$ $\frac{\partial (R^2)}{\partial t} + \sum_{k=1}^{N} \nabla_k \cdot \frac{R^2 \nabla_k S}{m_k} = 0$
2	$\vec{q}_k(t)$... position of an object at time t m_k ... mass of a particle S ... Hamilton Jacobi function or action R^2 ... probability Q ... Quantum potential Measurement rules for $\vec{q}_k(t), m_k, R^2$
3	Particle and wave conception Hidden variables Quantum potential
4	Novel matter ontology Active information Wholeness and ontological holism

Overview

We present the De Broglie-Bohm interpretation structure according to the Classification scheme in an overview. Differences from Bohmian mechanics, which will be presented in Sect. 2.2.3, are discussed in the conclusion (Table 2.4).

2.2.3 *Bohmian Mechanics*

Bohmian mechanics is a reconstruction of quantum mechanics based on the ideas of the De Broglie-Bohm interpretation, but postulates the guidance equation for particle positions and not as a consequence of the Schrödinger equation. It reproduces the results and experimental predictions of standard quantum mechanics, and is a realistic, deterministic, nonlocal hidden-variable theory. Its fundamental ontological entities are the particles and their position. We analyse Bohmian mechanics for N-particle systems without spin, because one-particle systems easily give a wrong impression on the nature of Bohmian mechanics. Main proponents of Bohmian mechanics are Detlef Dürr, Sheldon Goldstein and Nino Zanghi. For further reading, see Dürr and Teufel (2009), Cushing et al. (2013).

Level 1—The Mathematical Formalism

The mathematical level consists of two equations: the Schrödinger equation and the guidance equation. The function Ψ is a mapping from a $\mathbb{R}^{3N}$ vector space to the complex numbers together with the free parameter t.

$$\psi : \mathbb{R}^{3N} \times \mathbb{R} \to \mathbb{C} \tag{2.14}$$

$$(\vec{q}, t) \mapsto \psi(\vec{q}, t) \tag{2.15}$$

$$\vec{q} = \begin{pmatrix} \vec{q}_1 \\ \cdot \\ \cdot \\ \cdot \\ \vec{q}_N \end{pmatrix} \in \mathbb{R}^{3N} \tag{2.16}$$

$$i\hbar \frac{\partial \psi}{\partial t}(\vec{q}, t) = -\sum_{k=1}^{N} \frac{\hbar^2}{2m_k} \Delta_k \psi(\vec{q}, t) + V(\vec{q})\psi(\vec{q}, t) \tag{2.17}$$

$$\frac{d\vec{Q}_k}{dt} = \frac{\hbar}{m_k} \Im \frac{\nabla_k \psi}{\psi}(\vec{Q}, t) \tag{2.18}$$

Level 2—Physical Quantities and Connection to Empirical Rules

Many-particle Bohmian mechanics assigns properties to the kth particle,[4] its mass m_k, and its trajectory represented by a vector $\vec{Q}_k(t) \in \mathbb{R}^3$. This trajectory refers to the empirical accessible positions of particles[5] in time.

The laws of motion for N particles are formulated in a $\mathbb{R}^{3N}$ configuration space, where the complex valued ψ function is defined. The ψ function obeys Eq. (2.17). A solution can be provided when the function $V(\vec{q})$ is defined. $V(\vec{q})$ represents the physical potentials the particles are exposed to. The ψ function is called wave function.

The trajectories of the particles obey Eq. (2.18), which defines the trajectories due to the wave function, which is itself a solution of Eq. (2.17). The initial conditions of $Q_k(t)$ and $\frac{dQ_k(t)}{dt}$ have to be defined to find unique solutions for the trajectories.

The statistical distributions of quantum systems are reproduced by Bohmian mechanics; specifically, by the Quantum Equilibrium hypothesis (Dürr and Teufel 2009, p. 153):

> *Quantum Equilibrium Hypothesis. For an ensemble of identical systems, each having the wave function ψ, the typical empirical distribution of the configurations of the particles is given approximately by $\rho = |\psi|^2$. In short, Born's statistical law holds.*

This is a brief summary of the most important definitions and conceptions within Bohmian mechanics. Some aspects of the physical formalism presented here follow out of a principle, the already introduced Quantum Equilibrium hypothesis.

---

[4] A conception that will be defined on level three.

[5] In the sense of the empirical phenomena.

Level 3—Concepts and Principles

Quantum Equilibrium Hypothesis: The Quantum Equilibrium Hypothesis has the status of a concept or first principle in Bohmian mechanics.

The empirical statistical distributions of quantum systems are reproduced by Bohmian mechanics; specifically by the Quantum Equilibrium Hypothesis. It can be shown by Bohmian mechanics that if a quantum system's positions are distributed according to Born's statistical interpretation, then they will stay in this quantum equilibrium. The evolution of the ψ function maintains the quantum equilibrium (Dürr and Teufel 2009, p. 211ff).

Particle and wave conception: Particle: A particle is represented by a trajectory $\vec{Q}_k(t)$ and lives in the $\mathbb{R}^3$.

The wave assigned to quantum theory is represented by the wave function ψ and lives in the configuration space $\mathbb{R}^{3N}$.

Hidden variables: Bohmian mechanics is a hidden variable theory; these hidden variables are the particle's positions and their initial values.

Level 4—Ontology

Particle Ontology: Bohmian mechanics advances a strict particle ontology. The ontological objects are the particles, some of their properties and their positions (Dürr and Teufel 2009, p. 142):

> *Bohmian mechanics will ecplain that it is correct to say "is" for the positions of the systems particles, but that it is not correct to say "is" for other "observables".*

Dürr alludes to the fact that not all observables have an ontological status. In fact, position plays a distinguished role. Momentum or spin are not on the same ontological footing. They are not properties of the particles, the ontological objects themselves.

A further aspect of the particle ontology is that it is an object ontology. There are objects out there; these objects are the particles in a physical real space. Referring to the structuralist approaches where structures play an important role, moderate ontological structural realism or even more epistemological structural realism would be philosophical stances that are compatible with Bohmian mechanics.

All particles are guided by the wave function simultaneously, but this wave function lives in the configuration space, which has a dimension depending on the number of particles involved, or if we assume the wave function of the universe, it would have a dimension according to the number of particles in the universe. But it is not that easy to abandon the ontological footing of the wave function (Esfeld et al. 2013, p. 5–6):

> *There is indeed a prima facie good reason to do so and to admit the wave-function as a further concrete physical entity in addition to the particles.*

Mainly, the reason for this is grounded on the causal relation of the wave function to the movement of the particles. The aim for such a causal connection that the wave function is a guiding field led De Broglie to propose the original version of the guidance conception for the wave function. A further problem of this ontology would be Esfeld et al. (2013, p.7):

> *However, if one admits configuration space as a further stage of physical reality in addition to and independent of three-dimensional physical space, it is unclear how there could be a real connection between these two spaces that could amount to something existing in the one space guiding or piloting the motion of entities existing in the other one.*

The possibilities of solving this ontological connection range from neglecting the idea of any ontological footing of the wave function, or assigning the wave function to physical space, to the position that configuration space itself represents a fundamental ontology (Esfeld et al. 2013, p. 7).

Two positions seem promising and represent an ontological dualism within Bohmian mechanics: Humeanism and Dispositionalism (Esfeld et al. 2013, p. 25):

> *...Humeanism more precisely, Lewis thesis of Humean supervenience: the world according to quantum mechanics can be a vast mosaic of local matters of particular fact, namely the spatio-temporal distribution of the elements posed in the primitive ontology such as particle positions. The universal wave-function and the laws of quantum mechanics supervene on this distribution. They are nothing more than devices of economical bookkeeping, there being no real connections among the elements of the primitive ontology...*

Esfeld et al. (2013) emphasize that the mere fact of the existence of an approach like Bohmian mechanics refutes the view that quantum mechanics is in contradiction with Humeanism. In this view, the wave function is a instrumentalistic element of the theory without any footing in ontology. It is helpful to describe the situation but gives no deeper insight in the nature of reality. A classical analogy would be the center of mass of a bulk of matter. The center of mass conception is very helpful, but no ontological speciality is located at the position of the center of mass of the bulk of matter. It is just a nomological entity.

Dispositionalism, on the other hand, is the view that there is an inherent property of local beables, the particles, that ensures their evolution in a specific way. To Esfeld et al. (2013) it brings the advantage that:

> *... one can include something in the ontology that accounts for the temporal development of the local beables, without having to resort to accepting the wave-function in configuration space as an element of physical reality.*

Esfeld et al. (2013) favour dispositionalism but admit that it is an open research project to clarify the eligibility in foundation of physics:

> *On the question of Humeanism versus dispositionalism we side with dispositionalism, since we take it to be a sound demand to call for something in the ontology that accounts for the temporal development of the elements of physical reality and that grounds the law of motion, thus providing for real connections in nature.*

Nonlocality and Holism: Bohmian mechanics is a nonlocal theory and its ontology is also nonlocal and in some sense holistic (Dürr and Teufel 2009, p. 131):

Table 2.5 Classification scheme of Bohmian mechanics

1	$i\hbar\frac{\partial\psi}{\partial t}(\vec{q},t) = -\sum_{k=1}^{N}\frac{\hbar^2}{2m_k}\Delta_k\psi(\vec{q},t) + V(\vec{q})\psi(\vec{q},t)$ $\frac{d\vec{Q}_k}{dt} = \frac{\hbar}{m_k}\Im\frac{\nabla_k\psi}{\psi}(\vec{Q},t)$
2	$\vec{Q}_k(t)$... trajectory of the kth particle m_k ... of the kth particle $\psi(\vec{q},t)$ ... wave function in the $\mathbb{R}^{3N}$ $\rho = \|\psi\|^2$... Born's statistical interpretation Measurement rules for $Q_k(t), m_k, \rho$
3	Point particle concept Quantum Equilibrium Hypothesis
4	Particle ontology Position in $\mathbb{R}^3$ space and time Wave ontology: Humeanism versus Dispositionalism

> *What is new[6] is that the description of nature needs a function on the configuration space of all particles in the system. And why is that revolutionary? The point is that such a description involves all particles in the universe at once, whence all particles are "entangled" with each other,...*

The Nonlocality in Bohmian mechanics reflects an ontological holism as proposed in the De Broglie-Bohm interpretation, but Bohmian mechanics puts no emphasis on this aspect and takes it merely as a consequence of the mentioned dispositions of the evolution of the local beables.

Overview

We apply the classification scheme to give an overview on the structure of Bohmian mechanics. Significant differences concerning all levels of the classification scheme between the De Broglie-Bohm interpretation and Bohmian mechanics are evident (Table 2.5).

[6] In quantum mechanics compared to classical physics.

Chapter 3
Philosophy of Science from Artistotle and William of Ockham to Structuralism

3.1 William of Ockham's Philosophy in Historical Context

Since Ockham's razor is often used as a heuristic to argue in favour of one interpretation of quantum mechanics or another we want to clarify its origins, its philosophical embedding in philosophy of science and discuss its application in quantum mechanics. First, we provide an overview of the historical context and environment to understand the origins of the principle of Ockham's razor.

William of Ockham was a Franciscan friar who lived between 1287 and 1347. He received a theological education in the Franciscan order at the London convent, and later in Oxford, but he never completely finished with a master of theology, though he fulfilled all requirements. It is believed that this was due to his controversial views, which antagonized some of his fellows and superiors, and to some conflicts between the Franciscan order and Pope John XXII.

Following Beckmann (1995), Ockham's philosophy is pervaded by three major principles: The omnipotence principle, the contradiction principle, and the parsimony principle. All three of them have strong interrelations and a deep theological connection. This theological context is irrelavent to our application of Ockham's razor in physics, but is important to understand Ockham's epistemological ideas in the context they were developed.

It is important to note that these principles have never been set and classified by Ockham himself, and this classification is not used in literature universally. Nevertheless, these principles provide a good overview of the ideas of Ockham, which is the reason for us to make use of them here.[1]

[1] By analysing Ockham's original writings, it is clear that no such structure of principles is defined there. Beckmann (1995) emphasizes that as well.

G. C. Krizek, *The Conceptions of Reality in the Interpretations of Quantum Mechanics*,
https://doi.org/10.1007/978-3-031-65786-3_3

3.1.1 The Omnipotence Principle

The omnipotence principle describes the omnipotence of the theological entity god and its unrestricted ability to act as it wishes to do. In Ockham's own words (of Ockham 2012, OT VII, 45):

> *Deus nullus est debitor. or God is no one' debtor.*[2]

Following Beckmann (1995) Ockham's god is nearly unrestricted in his actions to create the world according to his will and free to change it as he wishes. Although the theological entity god can act completely unrestricted, there is one restriction: god is not able to act contradictory, and nothing in his creation is allowed to be contradictory. With this idea Ockham encountered the philosophy of necessitarianism, which had its origins in the work of Aristotle and Platon, where all things exist due to necessity and therefore our world as it exists is the only possible one. In a theological context, Ockham's position was radical and new, and was one of several reasons for Ockham's conflict with censorship and the papal court (Beckmann 1998; Klocker 1985).

As mentioned before, the restriction of the entity god is contradiction: this is settled in the second principle. For further details on this topic, refer to Schröcker (2003).

3.1.2 The Contradiction Principle

Ockham's contradiction principle is a continuation of aspects of the aristotelian theory of thought. Aristotle formulated this first principle in the following way (Ross 1928):

> *It is, that the same attribute cannot at the same time belong and not belong to the same subject and in the same respect.*

In Ockham's epistemology, the idea is continued in the sense that contradictions cannot even be purpose of knowledge and perception. According to Ockham, perceptibility and contradictoriness are fundamentally incompatible. The striking point is not that this idea is new, but that this is Ockham's starting point to aim for nominalism; this is his way to deal with the problem of universals.

As outlined before, the problem of universals is the question if whether general terms like the colour "red" have an ontological existence on their own. If this would be affirmed, "redness" would be a qualia on its own. The problem of universals has been discussed for centuries and continues to be a topic of debate even today. Famous participants in the discussion were Kant, Peirce, Russel, Whitehead and Wittgenstein (1889; 1974; 1912; 2003).

Ockham addresses the problem of universals in his comments on the "Four books of Sentences" by Peter Lombard, which was a standard exercise for every medieval

---

[2] Translation by the author.

theologian. The scholastic tradition dominated the medieval theology. It was based on the dialectical method, where a question is formulated, arguments against and in favour of a statement were discussed and replies to the objections were given. An analysis of this scholastic method, as an example by Ockham himself, can be found in Appendix A.

With his nominalist position Ockham argues that the objects of our perception in the world are individuals; they are unique and it is not necessary to use the concept of universals to describe these objects. Moreover, the universals are not entities that are necessarily needed; they are connected to the construction of language and the way how the human mind perceives and processes the world. The antithetic concept to nominalism is realism. It is important to distinguish this definition of realism, which only refers to the dispute on universals, from the realism concepts in philosophy of science such as naive realism or scientific realism. The realists position concerning the problem of universals assigns an objective existence to universals and goes back to the ideas of Platon. Therefore, the realist position is often called Platonism or Platonic realism, even though these concepts have several differences. Regarding the problem of universals, their position is identical; thus, we stick to this naming and subsequently refer to this position as Platonic realism.

Now we show how the contradiction principle plays an important role in Ockham's reasoning against Platonic realism.

Following Schönberger (1990, p. 105) Ockham's way of arguing is like this:

> *2 contradictory statements, if they should both be true at the same time, can only refer to two different things. And vice versa: Wherever something distinguishable occurs, contradiction can be claimed. Therefore a designated contradiction, if its contradictory contents can refer to distinguishable things, is a ground-breaking argument for Distinguishability or Difference.*

Or with Ockham's own words (of Ockham 2012, OTh II, Liber I, D.2 Q.6, p. 174):

> *contradictio est via potissima ad probandum distinctionem rerum.*
>
> *or*
>
> *Contradiction is the most important way to prove difference.*[3]

The contradiction principle should be understood as an attack on the concept of universals and an support of nominalism. Let's explore this further through a gedankenexperiment.

Assume a box as shown in Fig. 3.1, where a person examines its contents twice. Between these examinations, it should be possible for someone to secretly replace the objects in the box with different ones. The question is if whether the objects in the box at both times are recognized as identical. The objects we assume in our Gedankenexperiment have properties A and B; for convenient illustration, let it be colour and shape. Furthermore, we assume both properties A and B are dichotomic, so for colour there are only to possible states red or green; for shape let it be round or angled.

[3] Translation by the author.

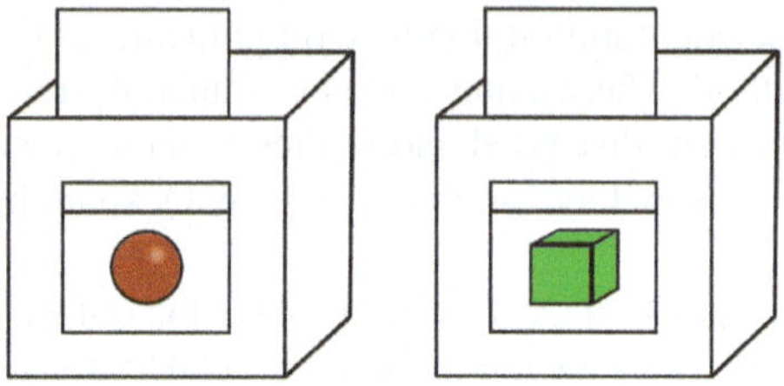

Fig. 3.1 The objects in the boxes

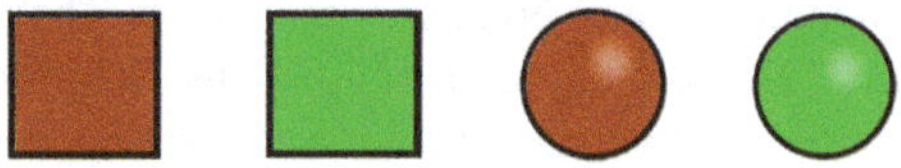

Fig. 3.2 The qualia of the objects

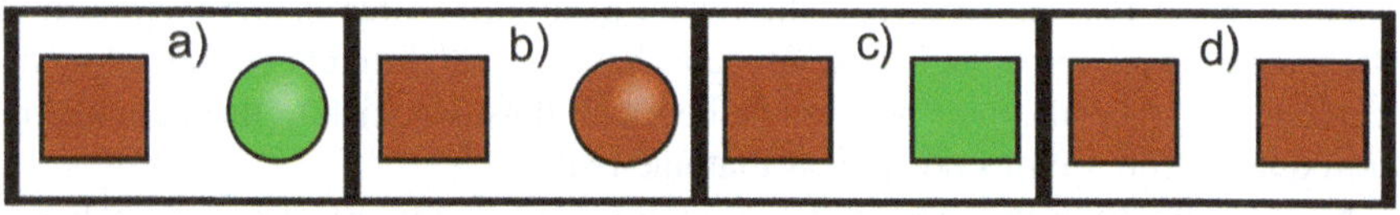

Fig. 3.3 4 of 16 possible combinations

Figure 3.2 shows the properties (qualia) of the assumed objects.

If the person controls the boxes twice, as defined, out of the 16 possible combinations, the possible cases, shown in Fig. 3.3, might occur:

(a)

$$colour_1 = red \quad colour_2 = green$$
$$shape_1 = angled \quad shape_2 = round$$

$$colour_1 \neq colour_2$$
$$shape_1 \neq shape_2$$
$$\hookrightarrow DIFFERENCE$$

(b)

$$colour_1 = red \quad colour_2 = red$$
$$shape_1 = angled \quad shape_2 = round$$

$$shape_1 \neq shape_2$$
$$\hookrightarrow DIFFERENCE$$

(c)

$$colour_1 = red \quad colour_2 = green$$
$$shape_1 = angled \quad shape_2 = angled$$

$$colour_1 \neq colour_2$$
$$\hookrightarrow DIFFERENCE$$

(d)

$$colour_1 = red \quad colour_2 = red$$
$$shape_1 = angled \quad shape_2 = angled$$

$$No \quad contradiction$$
$$\hookrightarrow IDENTITY$$

By the fact that there is a contradiction in the statements about the properties of the objects in the box at the two times, difference for the objects can be concluded. The conclusion is: Because of the contradiction, difference can be supposed. On the other hand, by perceiving no contradiction in the properties, it is reasonable to conclude identity for the objects.

In this last example, we avoided making use of universals; now let' introduce the concept of universals and add them to our gedankenexperiment. Additional to our dichotomic properties A and B, we assume that the universal "toybrick" should be applicable to all of our objects. It is important to understand that universals are not in general properties and, therefore, are not inherently perceptible to our observer.

(a)

$$colour_1 = red \quad colour_2 = green$$
$$shape_1 = angled \quad shape_2 = round$$
$$universal_1 = toybrick \quad universal_2 = toybrick$$

$$colour_1 \neq colour_2$$
$$shape_1 \neq shape_2$$
$$\hookrightarrow DIFFERENCE$$

(b)

$$
\begin{aligned}
colour_1 = red \quad & colour_2 = red \\
shape_1 = angled \quad & shape_2 = round \\
universal_1 = toybrick \quad & universal_2 = toybrick
\end{aligned}
$$

$$
\begin{gathered}
shape_1 \neq shape_2 \\
\hookrightarrow DIFFERENCE
\end{gathered}
$$

(c)

$$
\begin{aligned}
colour_1 = red \quad & colour_2 = green \\
shape_1 = angled \quad & shape_2 = angled \\
universal_1 = toybrick \quad & universal_2 = toybrick
\end{aligned}
$$

$$
\begin{gathered}
colour_1 \neq colour_2 \\
\hookrightarrow DIFFERENCE
\end{gathered}
$$

(d)

$$
\begin{aligned}
colour_1 = red \quad & colour_2 = red \\
shape_1 = angled \quad & shape_2 = angled \\
universal_1 = toybrick \quad & universal_2 = toybrick
\end{aligned}
$$

$$
\begin{gathered}
No \quad contradiction \\
\hookrightarrow IDENTITY
\end{gathered}
$$

As we see, adding the universal "toybrick" makes no difference to our conclusions on difference and identity. The reason is that all objects were attributed with the same universal. Let' assume that two toybrick manufacturers produce toybricks with our binary dichotomic properties A and B, colour (in red and green) and shape. Let them be LEGO© and K'NEX©. Now let us assume that the vicious person that wants to puzzle our experimenter, who is examining the properties of the objects in the boxes, always replaces a LEGO© toybrick at the first check with a K'NEX© toybrick for the second examination.

(a)

$$
\begin{aligned}
colour_1 = red \quad & colour_2 = green \\
shape_1 = angled \quad & shape_2 = round \\
universal_1 = LEGO© \quad & universal_2 = K'NEX©
\end{aligned}
$$

$$colour_1 \neq colour_2$$
$$shape_1 \neq shape_2$$
$$\hookrightarrow DIFFERENCE$$

(b)

$$colour_1 = red \quad colour_2 = red$$
$$shape_1 = angled \quad shape_2 = round$$
$$universal_1 = LEGO© \quad universal_2 = K'NEX©$$

$$shape_1 \neq shape_2$$
$$\hookrightarrow DIFFERENCE$$

(c)

$$colour_1 = red \quad colour_2 = green$$
$$shape_1 = angled \quad shape_2 = angled$$
$$universal_1 = LEGO© \quad universal_2 = K'NEX©$$

$$colour_1 \neq colour_2$$
$$\hookrightarrow DIFFERENCE$$

(d)

$$colour_1 = red \quad colour_2 = red$$
$$shape_1 = angled \quad shape_2 = angled$$
$$universal_1 = LEGO© \quad universal_2 = K'NEX©$$

$$No \quad contradiction$$
$$\hookrightarrow IDENTITY$$

We see, that in fact the universal assigned to an object with identical properties might be different, but there is no way to distinguish them. Therefore in Ockham's view it makes no sense to speak about the reality of those universals; they do not seem connected to the real world, if they do not reside in distinguishable properties of the objects. Furthermore, it would be possible to multiply them easily. If the toybrick belongs to Sarah, you could assign it a new universal, "Sarah' toybrick," and so on. This numberless amount of universals that would then reside in reality, in the view of Platonic Realism, was unnecessary to Ockham. It is not useful to introduce more entities than necessary.

This is the essence of Ockham's Parsimony principle, and one recognizes the strong connection to the Contradiction principle. In Ockham's own words of Ockham (2012, OP I, SL I, 12) or here von Ockham (1996, SL I, 12 p. 58):

Frustra fit per plura quod fieri potest per pauciora.

For nothing is done with a multiplicity, what can be done with less.[4]

We discuss the Parsimony principle in more detail in Sect. 3.1.3.

3.1.3 *The Parsimony Principle*

The parsimony principle, commonly known as Ockham's razor, is the heuristic method people refer to when they mention Ockham's razor. There are several formulations in the literature which are not originally by Ockham himself. Furthermore, it is not the case that there is a specific writing of Ockham where he states this principle as fundamental and assigns deeper relevance to it (although it is anchored in his whole thinking and pervades his philosophical and theological works). One original formulation can be found in of Ockham (2012, OP I, SL I, 12) or in von Ockham (1996, SL I, 12 p. 58):

Frustra fit per plura quod fieri potest per pauciora.

We may translate this sentence as

For nothing is done with a multiplicity, what can be done with less.[5]

or in more modern words

It is pointless to do with more what can be done with fewer.[6]

Several other formulations used in modern literature are not by Ockham himself. Thorburn (1918) clarified the question around the formulation and gives a comprehensive overview about formulations of the parsimony principle, but only in Latin. To illustrate the idea in its historical context, it makes sense to give translations[7]:

(a) *A multiplicity may never be stated without necessity.*

(b) *...multiplicity may not be stated, but for it is necessary .*

(c) *This opinion states a multiplicity without necessity, which is against the doctrine of philosophers.*

(d) *As somebody who follows the common sense does not state multiplicity, but for that what is included by common sense, somebody that follows Believe may not state more than the truth of Believe requires.*

[4] Translation by the author.

[5] Translation by the author.

[6] Translation by the author.

[7] Translation by Josef Reiter and the author.

(e) *The use of multiplicity must obviously always be necessary.*

(f) *For nothing is done with a multiplicity, what can be done with less.*

(g) *Stringency should be posed, where multiplicity is not necessary.*

From these passages and the context of the other principles we can conclude:

- As shown in Sect. 3.1.2, the contradiction principle plays an important role in Ockham's arguments in favour of his nominalism, countering the existence of the universals, which to him are numberless in principle. These universals represent the multiplicity he seeks to eliminate. This is the reason why he is applying the razor principle within this framework.
- There is no statement against the existence of entities in general, nor does the parsimony principle represent the view that entities must adhere to some extremum principle.
- The razor is no deep mystery to Ockham; it is just common sense. It is a very hands-on principle, a principle of economy of thought and resources.
- It is a principle of adequacy. Parsimony means not to reduce to a minimum at all costs. Parsimony means to use as much as necessary and as little as possible.

From this, we can conclude that Ockham's razor has no meaning as a strict ontological principle; it is not meant to be applied to ontological entities. It is an epistemological statement to minimize the number of symbols used to describe a certain sensational experience.

It is again Wittgenstein who sums up the razor principle (Wittgenstein 2003, 5.47321 and 3.328) in a comprehensible way:

> *Occam's maxim is, of course, not an arbitrary rule, nor one that is justified by its success in practice: its point is that unnecessary units in a sign-language mean nothing.*
>
> *If a sign is useless, it is meaningless. That is the point of Occam's maxim.*

To Ockham, the universals represent such useless symbols that do not reside in reality.

The notion of simplicity is strongly connected to Ockham's razor. It is often referred to as a principle of simplicity. That is somewhat misleading and should be discussed in more detail. Simplicity is a very subjective criterion unless it is somehow quantified by a measure of a sort of complexity. Often Ockham's razor is understood as the concept that only simple explanations are the correct ones; in the context of a theory, the simplest theory must be right. If that would be the case, in a naive way it would be pretty easy to do science, but this is obviously not the case. The simplicity goes only as far as possible. Adequacy and consensus with the experiment are favored over simplicity. Physics does not describe the world itself, it describes mental representations of experiences we conduct in the real world. Therefore, the simplicity, if applicable, does not belong to the world, but to the mental representation of the experiences. It is a comprehensible concept that human minds select the simplest way to describe an experience by mental representations. This concept is economical and parsimonious concerning the limited resources of the human mind,[8] but it does not necessarily tells us more about the things in reality. The success of science and its miracle, if you would like to use the notion of the miracle argument, is that our mental representations deliver models with predictive power.

Wittgenstein emphasizes the subjective element of the parsimonious method (Wittgenstein 2003, 6.363 and 6.3631):

> *The procedure of induction consists in accepting as true the simplest law that can be reconciled with our experiences.*
>
> *This procedure, however, has no logical justification but only a psychological one.*

Others emphasize the connection between simplicity and truth. Kelly (2011) formalises the concept of Ockham's razor for truth-finding and sets up an Ockham efficiency argument. A detailed analysis of this ideas would go beyond the scope of this work, anyhow the notion of truth is something difficult in context with physics and science in general. In some some technical applications and domains like machine learning a more applied notion of truth can be useful, but without any relation to the notion of truth in philosophy of science.

We will be concerned with the notion of simplicity again in the context of the interpretations of quantum mechanics, but finally we present an application of Ockham's razor to the domain of biology to demonstrate the attempt to apply Ockham's razor in other natural sciences as well. We argue that theses domains are not the one Ockham intended to apply Ockham's razor.

Westerhoff et al. (2009) argues in his justification for Systems Biology that

> *Two paradigms, i.e. Ockham's razor and the prevalence of minimum energy solutions are pertinent to much of physics*

Futher he is referring to living organisms:

> *For researching living organisms, Occams razor is not an appropriate paradigm.*

He argues that this is the case due to the high complexity and the high amount of constituents involved. He dismissive argues that you can speak about simplicity in the scale of ...

> *two or three components, and if that does not work perhaps 6. Three hundred and seventy-five is certainly not in the realm of simplicity.*

We want to emphasize again that the application of Ockham's razor in the regime of entities in reality[9] is not the intended application by Ockham. Ockham's arguments are applicable for mental representations that should fulfill the parsimony, not for physical systems or entities in reality. The application of Ockham's razor as a concept of parsimony or simplicity in the things itself goes far beyond the ideas of Ockham, therefore, it would be better to name efficiency concepts in different regimes other than that intended by Ockham, differently than "Ockham's razor".

---

[8] For further details refer to Tversky and Kahneman (1973).

[9] Like a biological system.

It would be an oversimplification to reduce Ockham's thinking to a few principles and fairly simple conclusions; it would not meet Ockham's philosophy in all its facets and subtleties, but for our purpose of a modern context it seems sufficient and appropriate to demonstrate his ideas in context of those principles. Further subtleties and the connection to the theological aspects[10] are discussed extensively in Schönberger (1990). For logical aspects of the contradiction principle, refer to Schick (2010).

3.2 Ockham's Razor and Other Heuristics

Before discussing the role of Ockham's razor in the interpretations of quantum mechanics, we provide an overview over other heuristics connected to Ockham's razor in recent physical debates.

We do not claim that the selection of heuristics presented here is comprehensive; instead, they have been chosen based on their relevance to our discussion.

3.2.1 Mach's Economy of Thought

Ernst Mach pleads for an idea of economy of thought, which emerged from his works in philosophy of science. He takes the point of view of phenomenology that all our scientific reasoning is based on economy of thought, starting with language, perception, and mathematics pervading all fields of scientific work. By perceiving sensations of things in the world, classifying them into boxes[11] and constructing laws out of their behavior, one does not gain knowledge about the things itself, one merely connects the mental representations of the things with laws (Mach 1898):

> *It thus comes to pass that we form the notion of a substance distinct from its attributes, of a thing-in-itself, whilst our sensations are regarded merely as symbols or indications of the properties of this thing-in-itself. But it would be much better to say that bodies or things are compendious mental symbols for groups of sensations, symbols that do not exist outside of thought. Thus, the merchant regards the labels of his boxes merely as indexes of their contents, and not the contrary. He invests their contents, not their labels, with real value.*

---

[10] The contradiction principle has one more aspect worth mentioning, which is negligible for the physical context of Ockham's razor, but still interesting for the historical embedding of Ockham's philosophy and theology. The contradiction principle is in connection with the omnipotence principle in the context that it even is impossible for the entity god to act in contradictory ways.

A modern account of this idea has been given by Wittgenstein (2003, 3.031):

> *It used to be said that God could create anything except what would be contrary to the laws of logic. The truth is that we could not say what an "illogical" world would look like.*

This principle was stressed mainly during disputes about Trinity, because of its inherent logical contradiction (Beckmann 1998).

[11] The boxes mean the categories in which the sensations of things are sorted.

His economy of thought relates to the mind-based processes where the laws interconnect the mental representations of the things, the mental symbols, in an economical way. So the economic character emerges from the perceiving subject and its inherent limitations.

It is an economy of practicality, even in mathematics (Mach 1898):

> *The greatest perfection of mental economy is attained in that science which has reached the highest formal development, and which is widely employed in physical inquiry, namely, in mathematics. Strange as it may sound, the power of mathematics rests upon its evasion of all unnecessary thought and on its wonderful saving of mental operations. Even those arrangement-signs which we call numbers are a system of marvellous simplicity and economy. ... No one will dispute me when I say that the most elementary as well as the highest mathematics are economically-ordered experiences of counting, put in forms ready for use.*

and further:

> *Physics is experience, arranged in economical order.*

It can be clearly seen that the economy Mach proposes is connected to the limited capacity or ability of the perceiving subject. It is therefore a practical economy justified by human beings as operators of scientific methods, not an economy settled in the things as they are. Economy is understood as a consequence of the development of predecessors of science, like handcraft or engineering, which made it necessary to transfer knowledge and experience. SI-units are human-centric from their choice of scale; likewise, the economic nature of physics resides in the economic nature of human language and thought.

For a more detailed analysis and different aspects of Machs economy of thought, refer to Kallfelz (1929).

3.2.2 *The No Alternative Argument*

The No Alternative Argument states that if despite intensive research, no alternative theories have been found to a specific theory or theory complex, this gives an argument in favour of the specific theory. It sounds simple, but has been put on a formal foundation in Dawid et al. (2014).

The argument of Dawid et al. (2014) refers to string theory to give justification in absence of empirical evidences. It is part of the claims on non-empirical evidences and is connected to the Meta Inductive Argument and the Unexpected Explanation Argument.

3.2.3 *The Meta Inductive Argument*

The Meta Inductive Argument says that if a specific theory has something in common with theories that have been empirically confirmed, it is reasonable to favor this specific theory (Dawid et al. 2014):

> *Now, assume that a novel theory H shows similarities to theories H1, H2, etc., in the same scientific research program. The joint feature of these theories may be a certain theoretical approach, a shared assumption, or any other relevant characteristic. Let us assume that a substantial share of the theories to which H is similar have been empirically confirmed. Assume further that for those theories, we have empirically grounded posterior beliefs about the number of alternatives. Then, it seems reasonable to use these posteriors as priors for the number of alternatives to H. After all, H is quite similar to H1, H2, etc. In statistics, such a way of grounding "objective" prior beliefs in past experience is referred to as the empirical Bayes method*

The Meta Inductive Argument is connected to the No Alternative Argument and Scientific Underdetermination, where it is argued that the number of possible alternative theories to a specific theory is limited in principle (Scientific Underdetermination Dawid 2007; Dawid 2008) and the No Alternative Argument favors the specific one if there is a lack of alternative theories. Furthermore for Dawid et al. (2014) the Meta Inductive Argument is not working standalone; it has to be seen in context with the Unexpected Explanation Argument.

3.2.4 *The Unexpected Explanation Argument*

The Unexpected Explanation Argument claims that a theory is more probable if the theory was found for a different domain but provides unexpected explanations (Dawid 2015):

> *Theory H was developed in order to solve a specific problem. Once H was developed, physicists found out that H also provides explanations with respect to a range of problems which to solve was not the initial aim of developing the theory.*

The Unexpected Explanation Argument is not an empirical argument, nor should it be mixed up with a novel empirical confirmation that supports a certain theory.

The idea of Non-Empirical Confirmations refers to the arguments presented in Sects. 3.2.2, 3.2.3 and 3.2.4 and uses them in combination to give justification to theories that bear an inherent lack of empirical data. The arguments have been designed for use in context with string theory, but Dawid et al. (2014) argue for application in archaeology and palaeontology as well.

3.2.5 The Ontological Coherence Argument

We propose a new heuristic that probably has been already been used implicitly throughout history of science, the Ontological Coherence Argument (OCA).

For this heuristic we assume that a theory consists of its mathematical apparatus, its physical interpretation by labels and names, its interpretation by concepts and principles and an ontological embedding. The details of the definition of these four levels of a physical theory have been defined in Chap. 2.

Theorem 3.1 *Under the assumption of a set of theories H, that cover different regimes of physics, and that H are Theories in agreement with experimental data in the respective regime, the OCA claims that those Theories $OT \in H$ whose ontologies fit together coherently are preferable (Fig. 3.4).*

The idea of OCA is not to rule out any discussed theory or interpretation within the framework of quantum mechanics, but to encourage to look out for connected theories outside of quantum mechanics to see if the puzzle of ontological entities gives rise to a bigger picture. A historical example where the application of the OCA can be seen applied is "The Great Cosmological Controversy" between the Geocentric, Heliocentric and Geoheliocentric model. Today it is widely unkown that there were striking empirical arguments in favour of the Geoheliocentric model by Tyho Brahe (Graney 2010). The empirical arguments were telescopic measurements of star disks, which supported the Geoheliocentric model by the assumption that stars should all have the same order of size. It turned out later that the observed star disk was an optical artefact, an airy disk caused by diffraction.

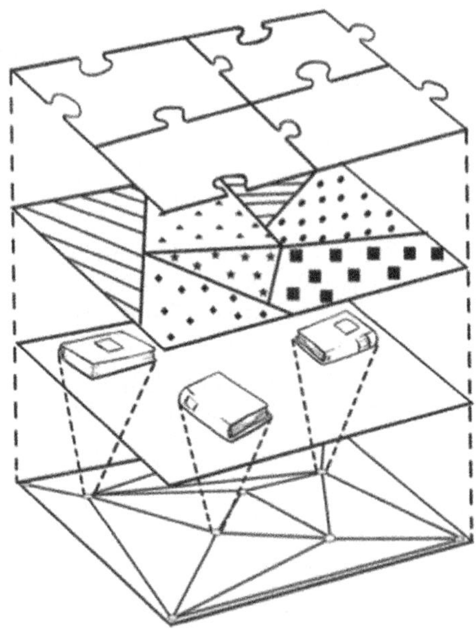

Fig. 3.4 Levels of interpretation

The picture of a new cosmology which connected to Isaac Newton' Principia Mathematica was ontologically striking; the Heliocentric view connected seamlessly to the nature of gravity and the laws of motion. Not only was it satisfying from the view of experimental confirmation, it was compatible through its ontologies and an implicit application of the OCA. For further details, refer to Kuhn (1957).

The OCA points in the direction of scientific realism, where the ontology of a theory plays a crucial role, and its connection to empiricism, instrumentalism, phenomenalism and constructivism will be investigated in a formal language in a subsequent work.

3.3 The Structure of Scientific Theories in the Framework of Philosophy of Science

Since many problems in the interpretations of quantum mechanics are connected with philosophy of science we provide an overview on certain aspects of it, specifically on the structure of scientific theories. Philosophy of science forms a framework for the scientific method, which evolved from early philosophy. Philosophy was and is pervaded by dualities, which will be of relevance in the discussion in interpetations of quantum mechanics. One of those foundational dualities is the contrast between Experience and Reason, or as (Einstein 1934) puts it:

> *I want now to glance for a moment at the development of the theoretical method, and while doing so especially to observe the relation of pure theory to the totality of the data of experience. Here is the eternal antithesis of the two inseparable constituents of human knowledge, Experience and Reason, within the sphere of physics.*

Another foundational duality is established by the realist and positivist positions. Within this framework, scientists' views on the construction of scientific theories have evolved in a complex manner over the last century, leading to a diversity of positions. We present an overview on those positions that are relevant to the discussion as they provide proposals on how to structure scientific theories. We present the positions and their relation to the classification scheme. The relevance of dualities will reappear in the context on interpretations of quantum mechanics, as different interpretations will have ontological commitments which are orthogonal in their positions.

3.3.1 Einstein's Model of Layers

Einstein presented his view on the structure of scientific theories in his work Einstein (1936):

> *Ziel der Wissenschaft ist erstens die möglichst vollständige begriffliche Erfassung und Verknüpfung der Sinneserlebnisse in ihrer ganzen Mannigfaltigkeit, zweitens aber die Erreichung dieses Zieles unter Verwendung eines Minimums von primären Begriffen und*

Relationen (Streben nach möglichster logischer Einheitlichkeit des Weltbildes bezw. logischer Einfachheit seiner Grundlagen). Die Wissenschaft braucht die ganze Mannigfaltigkeit der primären, d.h. unmittelbar mit Sinneserlebnissen verknüpften Begriffe sowie der sie verknüpfenden Sätze. In ihrem ersten Entwicklungsstadium enthält sie nichts weiter. Auch das Denken des Alltags begnügt sich im grossen Ganzen mit dieser Stufe. Diese kann aber einen wirklich wissenschaftlich eingestellten Geist nicht befriedigen, da die so gewinnbare Gesamtheit von Begriffen und Relationen der logischen Einheitlichkeit völlig entbehrt. Um diesem Mangel abzuhelfen, erfindet man ein begriffs- und relationsärmeres System, welches die primären Begriffe und Relationen der "ersten Schicht" als logisch abgeleitete Begriffe und Relationen enthält. Dieses neue "sekundäre System" erkauft die gewonnene höhere logische Einheitlichkeit mit dem Umstande, dass seine an den Anfang gestellten Begriffe (Begriffe der zweiten Schicht) nicht mehr unmittelbar mit Komplexen von Sinneserlebnissen verbunden sind. ... So geht es fort, bis wir zu einem System von denkbar grösster Einheitlichkeit und Begriffsarmut der logischen Grundlagen gelangt sind, das mit der Beschaffenheit des sinnlich Gegebenen vereinbar ist.

The goal of science is in the first place the complete conceptual acquisition and connection of the sensations in all its manifold aspects, secondly the attainment of this goal with a minimum of primary terms and relations. (ambition for a optimized logical unity of the world view, respectively logical simplicity of its foundations). Science needs the whole manifold of primary terms, i.e. which are directly connected to sensations, and the statements connecting them. In its first development stage it contains nothing more. Common sense thinking is satisfied with this stage. Though this stage cannot satisfy a truly scientifically oriented mind, since the resulted wholeness of conceptions and relations totally lacks in logical unity. To resolve this insufficiency one invents a system more economical in its use of conception and relations, which integrates the conceptions and relations of the "first layer" as logically derivable. This new "secondary system" gains the higher logical unity by the circumstance that its initial conceptions (conceptions of the second layer) are not directly related to the sensations any more. ... And so it goes forth, until we reached at a system of at most thinkable unity and parsimony of logical foundations, which is in agreement with the sensations.[12]

The difference of Einsteins model to the classification scheme presented in Chap. 2 is that Einstein's model of layers refers only to Level 1 and 2 of our classification scheme. The development of higher layers through the unification and definition of new theoretical terms and concepts extends the layer concept to Level 3 of the classification scheme, encompassing concepts and principles. It is a matter of the specific theory and its development to a maximally logical unified theory, if it reaches out to Level 3 of the classification scheme, or it is confined to the Levels 1 and 2. Einstein's layer model pictures the evolution of a physical theory from its first conceptions out of common sense, pure sensations and experience, to a physical theory with a maximum of parsimony in its used elements (Fig. 3.5).

It is surprising that within this layer model Einstein does not refer to an ontological level. Though the work is written in 1936, it relates to Einstein's early positions that are often seen in the positivist tradition of Mach. A detailed account on Einstein's alleged shift from positivism to realism is given by Howard (1993).

Einstein's model of layers is hierarchical and there is a maximum of logical unity a theory can develop to; it is represented by the highest layer in the model. The layer model is strongly influenced by the success in the evolution of Classical Mechanics

[12] Translation by the author.

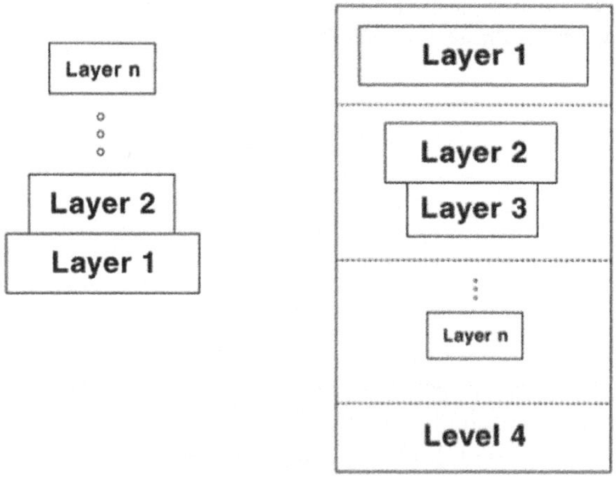

Fig. 3.5 Einstein's layers and the embedding to the classification scheme

from ancient conceptions to Newtonian Mechanics and General Theory of Relativity. It must have been this story of ongoing unification that motivated Einstein's view on theory structure.

This view is reflected also in the way Einstein presented the historical evolution of physics (Einstein et al. 1950).

3.3.2 The Received View

The received or syntactic view is the view on scientific theories that developed within the framework of Logical Empiricism in the 1920s in the environment of the Vienna Circle and influence of Machian positivism. A brief summary on the received view is provided by Suppe (1977, p. 12):

> *A scientific theory is to be axiomatized in mathematical logic (first-order predicate calculus with equality). The terms of the logical axiomatization are to be divided into three sorts: (1) logical and mathematical terms; (2) theoretical terms; and (3) observation terms which are given a phenomenal or observational interpretation.*

Between the theoretical terms and the observational terms, correspondence rules are defined. Every theoretical term has to have a corresponding observational term. Otherwise, it would not be an acceptable theoretical term within the received view. Instead, it would be considered a metaphysical term, having no relation to observations and, according to the stance of Logical Empiricism, can be disregarded.

These correspondence rules act between the Level 1 and 2 of the previously presented classification scheme. They do not refer to any metaphysical or ontological entity; they only represent relations between experience and theoretical terms.

$$Tx \equiv Ox \tag{3.1}$$

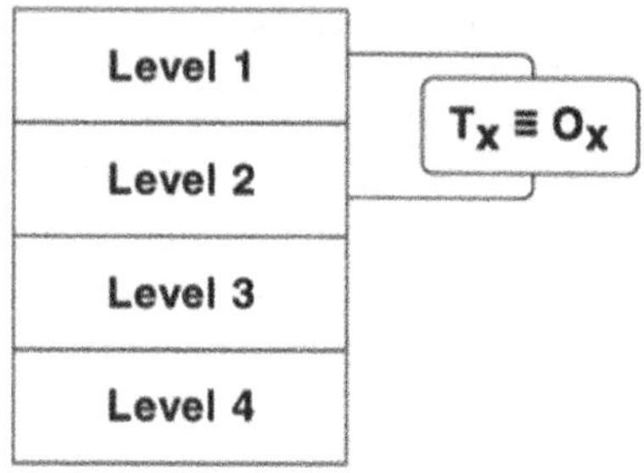

Fig. 3.6 The received view in the classification scheme

The received view failed due to several reasons, which we do not present in detail here. A detailed account is given by Suppe (2000), though two main reasons should be mentioned. Following Suppe (2000, p. 103), two critiques on the received view were essential:

> *theories are not linguistic entities and thus theories are individuated incorrectly.*

and (Fig. 3.6)

> *correspondence rules were a heterogeneous confusion of meaning relationships, experimental design, measurement, and causal relationships some of which are not properly parts of theories;*

The received view faced criticism, specifically concerning the position on the meaning of theoretical terms. As pointed out by Suppe (1977, p. 13), the summarized slogan of this position is

> *The meaning of a term is its method of verification.*

Carl Hempel, a developer and main proponent of the received view, changed his position during the Illinois Symposium on the Structure of Scientific Theories in 1969 and questioned one of the cornerstones of the received view (Hempel 1977, p. 253):

> *It presupposes that if theoretical terms have definite meanings, then it must be possible to construe those terms as introduced by specifiable logical procedures, which assign them meanings with the help of terms that are antecedently understood.*
>
> *This presupposition, which I once thought quite sound, seems to me mistaken for several reasons...*

This cleared the way for a new view on the structure of scientific theories, the semantic view, which developed out of the criticism on the received view. For an early presentation of the received view, refer to Carnap (1923); for detailed presentations of its failure, refer to Hempel (1970, 1977).

3.3.3 The Semantic View

The semantic view is a position that contrasted and succeeded the received view. It represents the view that a theory is represented by models, which makes use of mathematical terms and statements concerning theoretical terms.

Suppe (2000, p. 105).

The Semantic Conception identifies theories with certain kinds of abstract theory-structures, such as configurated state spaces, standing in mapping relations to phenomena. Theory structures and phenomena are referents of linguistic theory-formulations. The basic idea is that theory structures are identified with suitably connected families of models. Depending on mapping relationships required for theoretical adequacy, realist, quasi-realist or antirealist versions are obtained.

Whereas the received view neglected the need for metaphysical terms, the semantic view incorporates both possible polar positions of realism and antirealism on the question of ontological commitment (Suppe 2000, p. 106):

Such commitments are via individual mapping functions (Loc functions) from real-world objects to points in logical space.

and

a realist has Loc functions onto every state variable and maintains a theory is empirically true just in case theory-structure-allowed state transitions are identical to those possibly occurring in the actual world. Antirealists do not commit ontologically to all state variables. They only require countenancing Loc functions from observables and that theories be empirically adequate: If W is that portion of reality to which one attaches Loc functions, the image M of W is among the models comprising the theory.*

Within the semantic view, both positions can be represented, depending on which entities the Loc functions refer to. Represented in the classification scheme presented before, the realist and antirealist positions refer to the ontological level (Level 4) or to observables inside the theories framework (Fig. 3.7).

The semantic view itself has different versions, which have differences concerning the interpretation and role of confirmation (Da Costa and French 1990; Suppe 2000; Van Fraassen 1980).

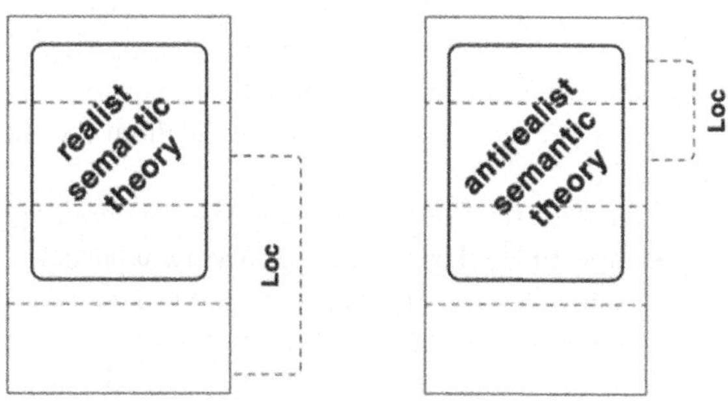

Fig. 3.7 Realist and antirealist positions within the semantic view

Out of the semantic view, the structuralist approaches in philosophy of science emerged.

3.3.4 *The Structuralist Approaches*

There are several accounts on structuralism. Three main approaches are presented in Schmidt (2014). We present them, and their relation to the classification scheme, in an overview.

An account on the development of the Structuralist Approach out of the Semantic approach is provided by Suppe (2000, p. 108):

> *Like the Semantic Conception it analyzes theories set-theoretically as comprised of a theory structure and intended applications, but is neo-positivistic in spirit and reliance on a relativized theoretical/nontheoretical term distinction. It began with Sneed's (1971) application of Suppes's (1957) set-theoretic techniques to the problem of theoretical terms.*

The structuralist approach by Joseph Sneed is a continuation of the positivistic end of the semantic view. It utilizes the set-theoretic methods to model a scientific theory. Hand in hand with the application of set-theory goes the neglection of linguistic terms (Sneed 1976, p. 144):

> *Roughly speaking the way of talking about scientific theories I am going to describe invites us to look at sets of "models" for these theories rather than the linguistic entities employed to characterize these models.*

Following the account in Schmidt (2014), we present the conception of Sneed's approach. A recent introduction to these basic structures in Sneed's structuralism approach and their relation is provided by Andreas and Zenker (2014).

According to Sneed (1976, p. 120), a scientific theory is a

> *... conceptual structure that can generate variety of empirical claims about a loosely specified, but not completely unspecified, range of applications.*

The structure is built by a core K and a domain of intended application I. The core contains the class of potential models M_p, which is basically the set of possible models that describe the situation in terms of the used elements in the model.

Assigned to the potential models, there is an actual model M that represents the set of potential models that satisfy empirical laws. Together, they form a model-element $\langle M_p, M \rangle$, the minimum element a theory has to contain.

M_{pp} is the set of potential models without the theoretical terms that have been involved in M_p.

As this M_{pp} set is difficult to understand without a practical application, we refer to the example of Hookes law for massive point particles attached to a spring, presented in Schmidt (2014). If M_p represents the full conceptual frame with the functions of force, the spring constants, the masses of the point particles and the position functions of the particles, then M_{pp} represents the model without all theoretical terms. This

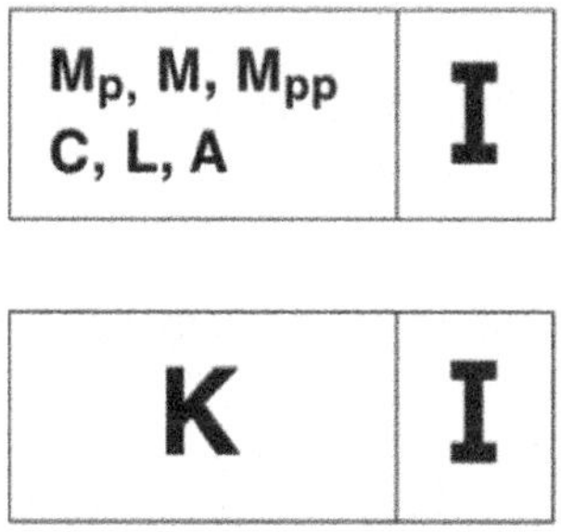

Fig. 3.8 The theory core K and the intended application I

would reduce the partial potential models to the particles positions as mass, spring constant and forces as theoretical terms.

Since the distinction between theoretical and non-theoretical terms is a peculiar question we refer to Andreas (2016) for an overview of positions and criticism.

Additional to M_p, M and M_{pp} the core contains constraints C that connect different models in the same theory, a class of links L that connect to models of other theories. For example, to describe the position of the point particles, an account on measuring distances in spacetime is needed.[13]

To give an account on the realistic situation of finite values in empirical sciences, a class of admissible blurs A is introduced, which contains degrees of approximation.[14]

The theory core K and the domain of intended application I form a theory element $\langle K, I\rangle$. The intended application I refers to the empirical data that the theory refers to; in fact, to the phenomena (Sneed 1976, p. 125):

> *The set I is to be interpreted as the range of intended applications of the element - what the theory-element is about. The only requirement (D4) puts on I is that its members have the structure characteristic of the non-theoretical part of K - that they be members of* $M_p p$ (Fig. 3.8).

Theory elements $\langle K, I\rangle$ form with other theory elements, a theory-net N. These theory-nets represent the complex we usually understand commonly as a scientific theory (Fig. 3.9).

A theory-net that undergoes a change over time describes a theory evolution I in time.

One essential concept in this structuralist approach is the extension process, which describes the reasoning from a limited empirical dataset represented by T-nontheoretical terms to T-theoretical terms (Gähde 2014, p. 1459–1460):

> *The starting point is provided by a certain set of data which is interpreted as a fragment of a model of an empirical theory. Attempts are made to extend this fragment into a complete model of that theory. ... this extension process is to be described by referring to the distinction between theoretical and non-theoretical functions with respect to the theory T in question. The intended applications of that theory (which comprise the available data) are located at*

---

[13] In the structuralist account of Ludwig, these links will be described by Pre-theories.

[14] In the structuralist account of Ludwig, these blurs A will be represented there by inaccuracy sets and unsharp measurements.

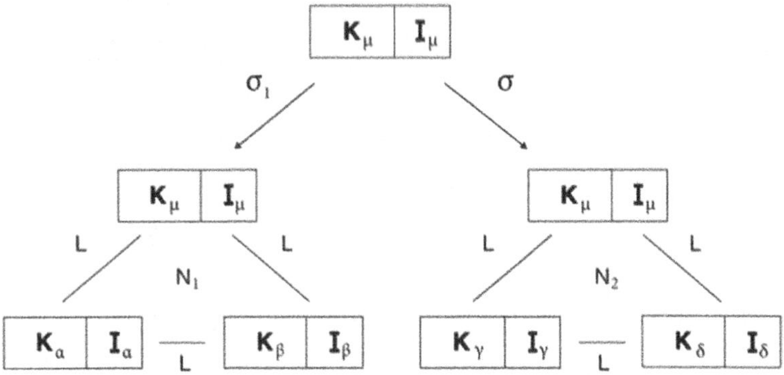

Fig. 3.9 The different elements of the structuralist approach

> *the non-theoretical level with respect to T and are represented by so-called partial models. These partial models consist of base sets and T-nontheoretical functions. By adding suitable T-theoretical functions, they are to be extended into models of T.*

Concerning the question on scientific progress, Sneed provides an instrumentalist answer (Sneed 1976, p. 143):

> *These considerations suggest that perhaps reduction of theories relative to specific nets is all we need to deal with the question of 'scientific progress'. They suggest that we do not need a concept of theory in which the possibilities for its future development are so strongly specified (through a strong specialization concept) that it can be established 'once and-for-all' that one theory can not keep up with another.*

Sneed's approach involves no reference to an ontology. It is a purely empirical account and is therefore confined to the Levels 1 and 2 of the presented classification scheme. Concepts and principles which are represented by Level 3 would contain linguistic terms and are neglected in the form of a statement by Sneed's structuralist approach. Principles like energy or momentum conservation are represented in the account, but not in the sense of a first order principle. They are merely empirical statements about and between the involved theory elements.

In contrast to Sneed's positivistic approach, a more realism-oriented approach has been presented by Günter Ludwig.

Günter Ludwig presented an approach based on the axiomatization of quantum mechanics embedded in an axiomatization of scientific theories in general. An overview is provided by Schmidt (2014):

> *His(G. Ludwig's) underlying philosophy is the view that there are real structures in the world which are pictured or represented, in an approximate fashion, by mathematical structures, symbolically $PT = W(-)MT$. The mathematical theory MT used in a physical theory PT contains as its core a species of structure Σ. This is a meta-mathematical concept of Bourbaki which Ludwig introduced into the structuralistic approach. The contact between MT to some domain of reality W is achieved by a set of correspondence principles $(-)$, which give rules for translating physical facts into certain mathematical statements called observational reports.*

Ludwig and Thurler (2007, p. 3)

> *The reality is in part constituted of facts stating "basic properties" of objects and "basic relations" between objects. Only facts related to the "domain of physics" are taken into consideration. These facts, directly recordable or indirectly recordable via known theories, called pre-theories, constitute what we call the physically recordable domain or the reality domain.*

Ludwig introduces the basic structure of a physical theory as the statement $PT = W(-)MT$. It is a correspondence between the domain of reality and a mathematical theory. In his last account Ludwig and Thurler (2007) altered the notion of the correspondence principle and replaced the usage to the symbol $(-)$, as a representation of the correspondence between reality and the mathematical theory, with the symbol (cor).

What Ludwig defines as an application domain is introduced by Einstein in his model of layers as "first layer" (Ludwig and Thurler 2007, p. 12):

> *The application domain of a particular physical theory PT_ν, denoted by $A_{p\nu}$ is the restriction of the reality domain W_ν to the facts that the theory considers a priori. The recording of facts can be made directly or indirectly by pre-theories.*

A pre-theory is a method to record those a priori given facts; it could, for example, be a pre-theory of distance measurement. The application domain is the collection of all those facts that build a basis for the theory PT_ν. The application domain of a theory can include reality domains W_α W_β of other physical theories PT_α, PT_β; they serve as pre-theories to the physical theory PT_ν.

The account of Ludwig contains one specific feature, the consideration of the finiteness of physics. He introduces inaccuracy sets to give regard to the fact that numbers of observations are finite, and the mathematical axiomatization is an idealization which involves an unavoidable inaccuracy.

This also reflects in the domains of physical reality. Additional to the application domain of a theory, he introduces a fundamental domain G_ν (Ludwig and Thurler 2007, p. 13):

> *The fundamental domain of a particular physical theory PT_ν, denoted by G_ν is the restriction of the application domain $A_{p\nu}$ to the facts that the theory describes. ... It is often more useful to apply the theory only on that part of the application domain $A_{p\nu}$ where we can use a small degree of inaccuracy. In such a region the theory essentially says something about the structure of reality and will be useful for technical application.*

If the theory PT_ν provides small inaccuracies over the whole application domain, the application and foundational domain are equivalent $G_\nu \equiv A_{p\nu}$.

At last there is a reality domain W_ν of the physical theory PT_ν (Ludwig and Thurler 2007, p. 14):

> *The reality domain of a particular physical theory PT_ν, denoted by W_ν, is the extension of the fundamental domain G_ν to the facts(related to the new physical concepts) that the theory describes. Our task is not only to detect "nonmeasured" realities, but also to detect new realities.*

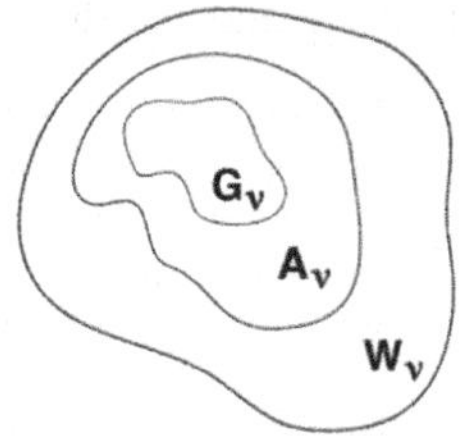

Fig. 3.10 The three domains of physical reality according to Ludwig

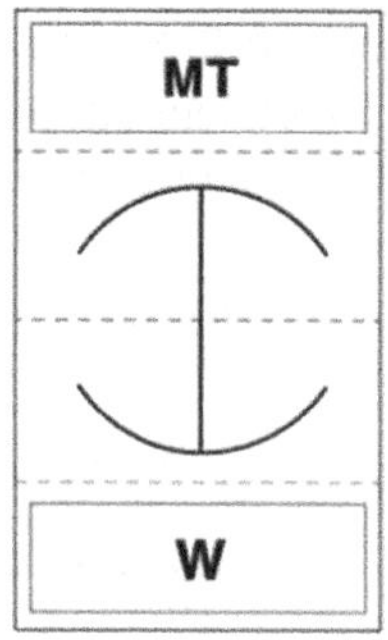

Fig. 3.11 The connection between Ludwig' structuralism approach and the classification scheme

The connection between the three domains of physical reality and the mathematical theory MT is shown in Fig. 3.10.

The three domains of physical reality contain each other as a subset (Fig. 3.11).

$$A_{p\nu} \subset G_{\nu} \subset W_{\nu} \tag{3.2}$$

The reality domain W_{ν} is corresponding, to our definition of the ontological Level 4 of a physical theory. The mathematical theory corresponds to the Level 1 and the correspondence rules (cor) between the mathematical theory and the domain of physical reality, which contain the interpretations and the concepts, refer to the Levels 2 and 3 of the classification scheme.

Ludwig's approach sides with realism on the spectrum of structuralist accounts.

The third structuralist approach we want to present is Erhard Scheibe's account and refers mainly to the mechanisms of reduction of theories. According to Schmidt (2014), his view can be seen between Sneed's and Ludwig's approaches. Although Scheibe's approach is not relevant to our discussion on a classification scheme, we mention his approach and its relation in the context of structuralist accounts. For an overview, refer to Scheibe (2001, 2013).

To summarize the idea of structuralist approaches, we refer to Balzer (1980, p. 399):

> *Die These des Strukturalismus läßt sich kurz so fassen: Erkenntnis der Wirklichkeit besteht darin, aus ihr Strukturen herauszulesen und in sie Strukturen hineinzuinterpretieren.*
>
> *The approach of structuralism explained in short: Insight about reality persits in perceiving the structures out of nature and interpreting structures into the realm of reality.*[15]

3.3.5 Epistemological and Ontic Structural Realism

Out of the structuralist ideas, new approaches evolved. Structural realism emphasizes the importance of the structural relations and assigns to them central relevance. The ontological role of objects in structural realism is a matter of dispute, and the different positions within structural realism assign different weight to the concept of reality of objects (Ladyman 2016):

> *The structuralist solution ... is to give up the attempt to learn about the nature of unobservable entities from science. The metaphysical import of successful scientific theories consists in their giving correct descriptions of the structure of the world. Theories can be very different and yet share all kinds of structure. The task of providing an adequate theory of approximate truth that fits the history of science and directly addresses the problem of ontological continuity has hitherto defeated realists, but a much more tractable problem is to display the structural commonalities between different theories.*

This abandoning of unobservable entities varies in extent. Within the structural realism, the bandwidth goes from a a realistic-oriented epistemological structural realism (ESR) to a positivistic-oriented position of ontic structural realism (OSR).

In the framework of the discussion of realist positions and anti-realist positions on the meaning of scientific progress and the value of ontological statements, three main arguments are noteable: The miracle argument by Putnam (1975) defending the position of realism, the Scientific Underdetermination (Stanford 2016) and the Pessimistic Meta Induction by Laudan (1981) on the side of anti-realism.

John Worrall attempted to find a synthesis of these ideas and asked if it is possible to get the best of both worlds (Worrall 1989, p. 101):

> *The main interest in the problem of scientific realism lies, I think, in the fact that these two persuasive arguments appear to pull in opposite directions; one seems to speak for realism and the other against it; yet a really satisfactory position would need to have both arguments on its side. The concern of the present paper is to investigate this tension between the two arguments and to suggest (no more) that an old and hitherto mostly neglected position may offer the best hope of reconciling the two.*

The ESR attempts to resolve the miracle argument by asserting that the commonality between an outdated theory and its successor lies in the relationships between their elements, i.e., their structures. In the outdated theory, reality has been described

---

[15] Translation by the author.

accurately enough, at least in a specific regime, so that the old description was eligible. The successor theory extended the application domain and incorporated the outdated old theory. The ontological commitments might have changed while proceeding to the new theory; the old theory might have introduced ontological entities that are superfluous in the successor theory, but the structures between objects are common in both theories. Therefore, ESR gives attention to these structures, outlined in an example on the theory of light (Worrall 1989, p. 117):

> *There was an important element of continuity in the shift from Fresnel to Maxwell - and this was much more than a simple question of carrying over the successful empirical content into the new theory. ... There was continuity or accumulation in the shift, but the continuity is one of form or structure, not of content.*

Concerning metaphysical concepts, ESR takes the position that metaphysical statements have to follow our best theories and are not to be seen as first-order principles (Worrall 1989, p. 123):

> *The only claim is that ultimately evidence leads the way: if, despite all efforts, no scientific theory can be constructed which incorporates our favourite metaphysical assumptions, then no matter how firmly entrenched those principles might be, and no matter how fruitful they may have proved in the past, they must ultimately be given up.*

Emphasizing the structures as the elements of a theory that ensure continuity in the shift from an outdated theory to a successor theory does not mean that ESR neglects the objects or entities on the ontological level, as those elements are connected by structures. Concerning mappings from theory elements to the ontological level, ESR advances a position considerably different than structuralist approaches before (Worrall 2007, p. 154):

> *But there is no reason why the way in which a theory mirrors reality should be the usual term-by-term mapping described by traditional semantics.*

In the context of the classification scheme, ESR could be seen as giving credit to a reality for objects, but not in an accessible way. The structures correspond to elements we can access, but there is more on the ontological level, the relata connected by the structures. These elements referring to the objects are epistemologically hidden from us (Fig. 3.12).

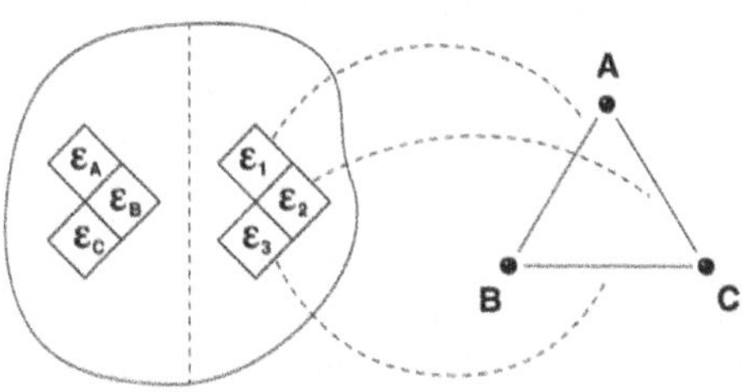

Fig. 3.12 ESR in the classification scheme

And further concerning this mapping, (Worrall 2007, p. 154):

> *SSR*[16] *in fact takes it that the mathematical structure of a theory may globally reflect reality without each of its components necessarily referring to a separate item of that reality.*

This aspect involves a specific form of holism in the conception of ESR that we cannot discuss in further detail here, but the relation to other holistic conceptions such as (Bohm et al. 1987; Quine 1951) and for a recent review on confirmational holism (Carlson 2015), is an open question.

ESR holds a scientific realism that denies that scientific theories provide access to these elements of reality at all Worrall (2007):

> *But why should a realist not be equally as fallibilist and tentative about the mode of reference of the terms in theories as she is about those theories truth? In any event, no one seriously holds, as I have remarked several times, that we have any theory-independent access to the furniture of the universe that would allow us to compare (even in principle) the notions conjectured by our theories with what there really is.*

To proponents of ESR, this inaccessibility is not an argument in favour of anti-realism. In contrast to ESR, Ontic structural realism has been introduced as a positivistic point of view on the structural realist account.

The origin of the ontic structural realism (OSR) is the distinction between epistemological and ontic structuralist approaches provided by Ladyman (1998a). OSR as well as ESR emphasize the role of structures as central elements of ontology (Ladyman 2016):

> *Ontic structural realists argue that what we have learned from contemporary physics is that the nature of space, time and matter are not compatible with standard metaphysical views about the ontological relationship between individuals, intrinsic properties and relations. On the broadest construal OSR is any form of structural realism based on an ontological or metaphysical thesis that inflates the ontological priority of structure and relations.*

The essential novelty of OSR is neglecting the individual objects and their properties as metaphysical and superfluous. Therefore, Ladyman's position is often referred to as radical OSR. A thorough review of the various views and peculiarities concerning the difference between epistemological and ontic structural realism is provided by Ladyman (2016).

It is notable that OSR is a realist account, but confined to the structures. If we identify the empirical content of the positivistic approaches by Sneed with the structures in OSR, OSR could be seen in the positivistic tradition of Logical Empircism. Then, OSR is a positivistic account concerning properties of objects and individuality.

In the classification scheme, the OSR approach would refer to the ontological Level 4, but only regarding the structures. Therefore, the identification of elements of reality is not due to properties of objects; it refers to structures and relations (Fig. 3.13).

One account that could be thought of as intermediate between ESR and radical OSR is advocated by Esfeld and Lam (2008). Moderate OSR holds the position that

---

[16] Structural Scientific Realism, is the wording John Worral assigns to this program by himself. Ladyman introduced ESR.

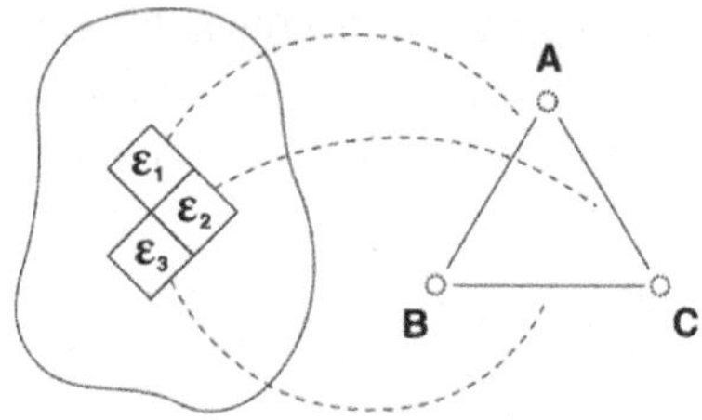

Fig. 3.13 OSR in the classification scheme

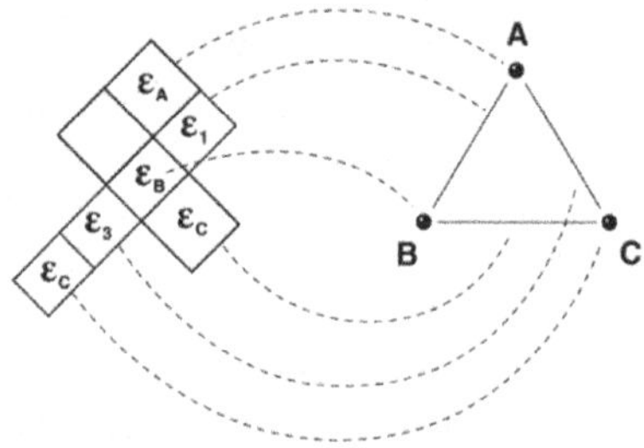

Fig. 3.14 Moderate OSR in the classification scheme

relations without relata are inconceivable. Challenging the radical version, Esfeld and Lam (2008, p. 29) states:

> *If it is claimed that there is something that exists but that we cannot know, we need an argument why we should accept that there is any such thing. The master argument for intrinsic properties can be summed up in this way:*
>
> *(1) Relations require relata, that is, objects that stand in the relations.*
>
> *(2) These objects have to be something in themselves, that is, they necessarily have some intrinsic properties over and above the relations that they bear to one another even if the relations do not supervene on the intrinsic properties and even if we cannot know the intrinsic properties.*

Moderate OSR takes a different position concerning the ontological status of the objects and the relation. Following moderate OSR, they are ontologically on the same level; none is primary (Fig. 3.14).

In the context of the presented classification scheme, this implies that there exists elements of reality for the properties and the structures on the same footing.

3.3.6 Quine-Duhem Thesis

The Quine-Duhem thesis is an approach that understands science as a system of statements that are connected with each other. There are statements that are far from experience; they form the center of a imagined sphere. Those statements involve logic and mathematics. Located on the outer regions of the sphere, connected to the

Fig. 3.15 The web of belief

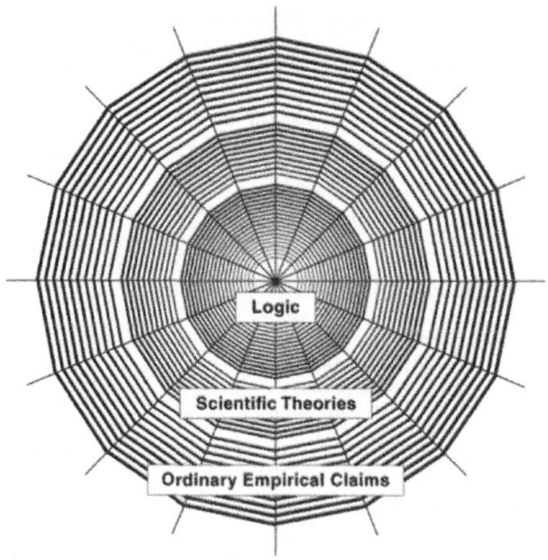

realm of mathematics and logic, are those statements that are in connection with experience. The scientific enterprise is represented by the structure of this web of belief and its development (Fig. 3.15).

If science and the network of statements that forms our web of belief gets in contradiction with experience, the whole web of belief is in contradiction. The Quine-Duhem thesis will state the following:

It is equivalent, in our intention, to remove the contradiction if we add or modify a single statement on the outer shores of the web of belief, or if we modify statements at the outermost heart of our system and thereby completely change the whole web of belief (Quine 1951).

This concept is called confirmational holism; it asserts that we never perform empirical tests in isolation. According to Popper, it challenges falsification because falsification would also pertain to the entire web of belief. IIn line with the principle of parsimony, it is beneficial to aim for the minimum necessary modifications during a contradiction. This stance is called Conservatism and is connected to the social psychology aspect of the scientific enterprise (Stanford 2016, Chap. 2.1); keep as much as possible, introduce as much as necessary.

An application for the Quine-Duhem thesis is the interpretation of quantum mechanics. Although there is no contradiction with experience, the interpretations of quantum mechanics demonstrate a situation where different sets of statements at the shores of the web of belief seem to be equivalent. The Quine-Duhem thesis would see those interpretations as equivalent, and there are even interpretations that demonstrate the change at the very heart of the system. Quantum Logic is an approach in the interpretation of quantum mechanics that modifies the statements in the realm of logic. For an introduction, refer to Mittelstaedt (1978) or Van Fraassen (1980).

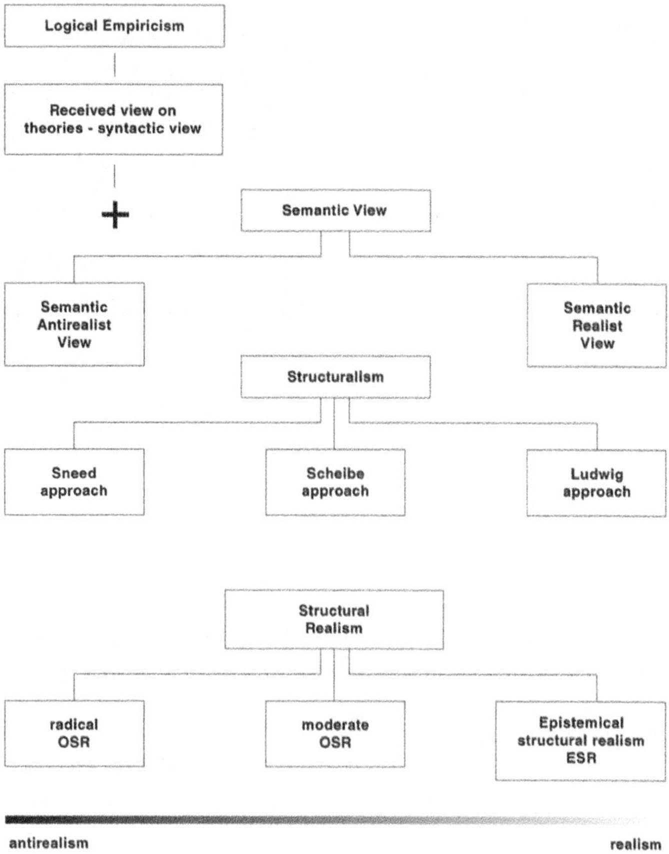

Fig. 3.16 An overview on the discussed points of view with respect to realism and antirealism

With this overview of the discussion on the structure of scientific theories within the philosophy of science, we return to the core problems in the interpretation of quantum mechanics (Fig. 3.16). In this regard, the debate on the foundations of quantum mechanics resulted in an influential work by Einstein, Podolsky, and Rosen in 1935, which continues to influence discussions to this day.

Chapter 4
The EPR Debate, the ER Connection, and Completeness

The year 1935 holds significant importance in the history of quantum mechanics. During this year, Albert Einstein, along with his collaborators Nathan Rosen and Boris Podolsky, published a seminal paper that challenged the completeness of quantum mechanics. Together with Nathan Rosen Einstein wrote a further paper on specific solutions of the field equations in general relativity that will later be named Einstein-Rosen bridges. Bohr will answer in the same year on Einstein objections on completeness. Erwin Schrödinger will correspond with Einstein on the matter and will be inspired to formulate his famous cat paradox. We discuss how some aspects of these events are connected and their relation to the interpretations of quantum mechanics.

4.1 Einstein's View on Theories and Theory Formation

In 1934 Einstein was driven by the long-lasting debates on the foundations of quantum mechanics to clarify his views on the method of theoretical physics in general. He emphasized the nature of the constituents of a theory used in physics (Einstein 1934):

> *For to the discoverer in that field, the constructions of his imagination appear so necessary and so natural that he is apt to treat them not as the creations of his thoughts but as given realities.*

and

> *He is in just the same plight as the historian, who also, even though unconsciously, disposes events of the past around ideals that he has formed about human society.*

This already indicates that the objection that Einstein was a proponent of a naive realism[1] has to be reconsidered. His ontology is, of course, not at trivial one, and we have to define what is meant exactly by naive realism. We will come back later

[1] See Janssen and Lehner (2014, p. 309ff).

G. C. Krizek, *The Conceptions of Reality in the Interpretations of Quantum Mechanics*,
https://doi.org/10.1007/978-3-031-65786-3_4

to that discussion. Einstein then highlights the second fundamental principle upon which science is based: its reliance on experience and the necessity for empirical verification:

> *Pure logical thinking can give us no knowledge whatsoever of the world of experience; all knowledge about reality begins with experience and terminates in it.*

It is remarkable that Einstein proposed his views parallel to Popper (2005), and they agree on several points. As we will come to understand later, their viewpoints are intricately interconnected. What are those ideals that Einstein is referring to? In his view, a theory is built upon concepts and laws and must adhere to certain requirements:

> *The basic concepts and laws which are not logically further reducible constitute the indispensable and not rationally deducible part of the theory. It can scarcely be denied that the supreme goal of all theory is to make the irreducible basic elements as simple and as few as possible without having to surrender the adequate representation of a single datum of experience*

Einstein is referring to Ockham's razor principle in this statement on the economy that the theory building process has to comply with. A detailed analysis of Ockham's razor in context to quantum mechanics is provided in Sect. 5.4.

Einstein's ideas on theory building are guided by the structure of the general Theory of Relativity and its success. To Einstein, it was representative to show the incorrectness of the induction principle as a theory building process and, to the contrary, consolidated his beliefs in a more top-down-orientated theory building process out of principles.

By rejecting induction as theory building principle and taking up the position that the elements of theories are creations of human thoughts, the question arises if there is a correct way to find insights about nature. And here Einstein makes clear that to him there is such a way (Einstein 1934):

> *To this I answer with complete assurance, that in my opinion there is the correct path and, moreover, that it is in our power to find it. Our experience up to date justifies us in feeling sure that in Nature is actualized the ideal of mathematical simplicity. It is my conviction that pure mathematical construction enables us to discover the concepts and the laws connecting them which give us the key to the understanding of the phenomena of Nature.*

His argument in favour of this position is the huge success of mathematics as a tool to describe our experiences about nature. It is this unreasonable effectiveness of mathematics that led Eugene Wigner to publish his paper Wigner (1960) of the same title. On the other hand, Einstein identified a problem in the above argument, when he stated that this mentioned success is claimed as well by theories that cannot provide a deeper ontological insight, such as classical mechanics.

> *Have we any right to hope that experience will guide us aright, when there are theories (like classical mechanics) which agree with experience to a very great extent, even without comprehending the subject in its depths?*

Einstein ignores this objection and heads for a worldview with a distinguished position of mathematics[2]:

> *...the truly creative principle resides in mathematics. In a certain sense, therefore, I hold it to be true that pure thought is competent to comprehend the real, as the ancients dreamed.*

It reflects a kind of mathematical realism, which is in one aspect contradictory, when Einstein pointed out that the elements of the theory are ideals, are pure constructions of human thoughts, but on the other hand represent a "correct way," reflecting a truth about nature.

As a basic entity of this worldview, he adopts the four dimensional continuum, which already was extremely successful for unification of physics during the early 20th century. He refers to this success and the principle of simplicity and states:

> *It is essential for our point of view that we can arrive at these constructions and the laws relating them one with another by adhering to the principle of searching for the mathematically simplest concepts and their connections. In the paucity of the mathematically existent simple field-types and of the relations between them, lies the justification for the theorist's hope that he may comprehend reality in its depths.*

Einstein presented and clarified his ideas on theory formation in general, about his ontological ideas and the scientific method in general. He presented field theories as candidates for this simple new theory that describes the real. His concern was mostly about how to integrate the atomistic structure of matter and energy.

We will see later that he used this wording again in his 1935 papers, and it is relevant to see what he meant exactly:

> *...the theory*[3] *in its basic principles is not an atomic one in so far as it operates exclusively with continuous functions of space, in contrast to classical mechanics whose most important feature, the material point, squares with the atomistic structure of matter.*

What Einstein understands as "atomic theory" or "atomistic structure" is discrete localization in space, in contrast to continuous functions of fields in space. And he emphasizes the importance of this idea in the conclusion of the paper:

> *Thus in a continuum theory, the atomistic character could be satisfactorily expressed by integral propositions without localizing the particles which constitute the atomistic system. Only if this sort of representation of the atomistic structure be obtained could I regard the quantum problem within the framework of a continuum theory as solved.*

To Einstein, this concept is of great importance to represent particles in a field theory. The singularity-free localization of particles is a core element of a successful theory. Some aspects of Einstein's views on the structure of theoretical physics directly relate to the nature of reality. Now we analyse the notions of Einstein's reality and the structure of theories.

---

[2] Wigner will refer to Einstein' objection 25 years later.

[3] He refers to the field theory.

4.2 Einstein's Not-Naive Realism and the Notions of Reality

We have presented a classification scheme to categorize different aspects of a theory. Therefore, we defined graphical representations of physical quantities, physical laws, physical concepts and elements of reality (Table 4.1).

Out of these icons, we can construct a depiction for a theory and for elements of reality inside objective reality (Fig. 4.1).

Table 4.1 Icons used in the classification scheme

	Physical quantity
	Law between physical quantities
	Concept connecting laws and physical quantities
	Elements of objective reality

Fig. 4.1 Elements of reality and the structure of a theory

Formally, we define a set of elements of a theory T_{PR} and a set of elements of reality PR.

$$T_{PR} = \{t_i\} = \{A, B, C\} \tag{4.1}$$

$$PR = \{\epsilon_i\} = \{\epsilon_A, \epsilon_B, \epsilon_C\} \tag{4.2}$$

We make use of them in the analysis of the reality criterion and the conceptions of completeness.

For classification, we proposed four levels in a theory to classify physical theories regarding their different aspects. The intention is to compare different interpretations of quantum mechanics regarding their formalism, concepts and ontologies.

- The first level contains the mathematical formalism of the theory, mathematical symbols and the connecting laws. The mathematical symbols can be identified with symbols for physical quantities, so we speak of physical quantities, although we mean only their symbolic representations.
- The second level contains the assignment of the previously defined physical quantities to sensations. Physical quantities consist of quantities and units, and need a procedure to be measured. All these assignments are contained in the second level.
- The third level contains concepts and principles that give reason and structure to the theory. First principles, such as extremal principles or conservation principles, belong to this level.
- The fourth level is the ontological embedding of the theory; it defines how physical quantities and concepts are related to ontological elements. Even if there are no ontological elements, such as in a positivistic philosophy of science, this would be defined here. Here, the elements of reality Einstein refers to are located. They are parts of the objective reality.

We use the classification scheme for the discussion of the concept of completeness, the criterion of reality, and in the analysis of naive realism in the Sect. 4.2.1.

4.2.1 Naive Realism

Naive realism, or direct realism, is the view that the perception delivers a full mapping between the elements of reality and the sensations. There are a lot of subtleties that could be discussed in that regard; some views of naive realism will require the spatiotemporal reference frame, some views will assume materialistic ideas. We confine ourselves to those aspects that are of interest for our discussion.

We already formulated the objection that Einstein was a proponent of naive realism. The term "naive" is depreciative inherently and includes prejudices, though will we make use of it due to the historical account, but we distance ourselves from a depreciative intent of this term. To get a better idea about the concept of naive realism, we present the definition by Russell (1940):

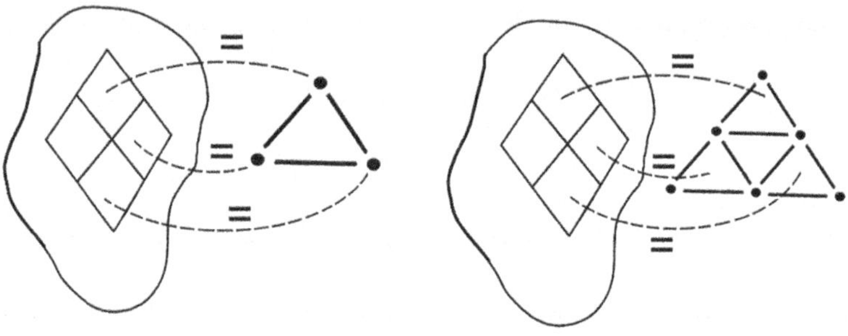

Fig. 4.2 Naive realism and an ontology for quantities and concepts

We all start from naive realism, i.e., the doctrine that things are what they seem. We think that grass is green, that stones are hard, and that snow is cold. But physics assures us that the greenness of grass, the hardness of stones, and the coldness of snow are not the greenness, hardness, and coldness that we know in our own experience, but something very different. The observer, when he seems to himself to be observing a stone, is really, if physics is to be believed, observing the effects of the stone upon himself.

Another statement of naive realism by Einstein (1946) shows Einstein's position on this matter (Fig. 4.2):

...naive realism,... according to which things "are" as they are perceived by us through the senses. This illusion dominates the daily life of men and of animals; it is also the point of departure in all of the sciences, especially of the natural sciences. ... The overcoming of naive realism has been relatively simple.

This last remark in Einstein (1946) shows that Einstein would not consider himself as a proponent of naive realism.

If we apply the classification scheme presented before on a definition of naive realism, one would conclude that there are identities between the physical quantities, or the physical concepts with the ontological elements of reality.

In the context of materialism, this would mean that the rigid bodies we experience through sensations and we model as a concept within Newtonian mechanics, would be assumed to exist as experienced. This view is rejected by physics and philosophy of science. A detailed discussion on the rejection of direct and naive realism can be found in Le Morvan (2004).

But still, a kind of naive realism is discussed in recent physics, the mathematical universe hypothesis (MUH) (Tegmark 2004):

Let us now digest the idea that physical world (specifically, the Level III multiverse) is a mathematical structure. Although traditionally taken for granted by many theoretical physicists, this is a deep and far-reaching notion. It means that mathematical equations describe not merely some limited aspects of the physical world, but all aspects of it. It means that there is some mathematical structure that is what mathematicians call isomorphic (and hence equivalent) to our physical world, with each physical entity having a unique counterpart in the mathematical structure and vice versa.

It is not our goal to discuss the MUH here, but we have seen before that Einstein had in mind a kind of mathematical realism. We discuss the notions of Einstein's realism later, after clarifying some subtleties in the EPR paper.

4.2.2 *Subtleties in the EPR Paper*

When we discuss the notion of reality, we have to clarify the used terms, and there we immediately face one subtlety. In the handwritten note for the talk he held in Oxford at the Herbert Spencer lecture on the method of theoretical physics, Einstein (1933) made use of the two synonymical terms "Wirklichkeit" and "Realität". In the English translation and published paper Einstein (1934) both terms merge to the term "reality." It is clear from the context in which Einstein made use of these terms that "Wirklichkeit" stands for the objectively given facts in nature, independent of observation. "Wirklichkeit" refers to the ontology in the world. He makes use of "Realität"only once in the context of an allegory to explain his personal struggle with philosophy of science (Einstein 1934):

> *For to the discoverer in that field,*[4] *the constructions of his imagination appear so necessary and so natural that he is apt to treat them not as the creations of his thoughts but as given realities.*

In Einstein et al. (1935), we find three wordings, but the paper was written originally in English: "objective reality," "physical reality" and "reality." The latter is just used as a short form for "physical reality". The definition of "objective reality" can be found in the introduction of the paper:

> *Any serious consideration of a physical theory must take into account the distinction between the objective reality, which is independent of any theory, and the physical concepts with which the theory operates.*

It is noticeable that Einstein defines the term "objective reality" and then in the definition of completeness makes use of "physical reality" (see citation in Sect. 4.3.3), which has not been defined before Einstein et al. (1935):

> *Whatever the meaning assigned to the term complete, the following requirement for a complete theory seems to be a necessary one: every element of the physical reality must have a counterpart in the physical theory*

Out of how the reality criterion refers to this "physical reality" and the elements of reality, and the lack of further definition of "physical reality" we conclude that "physical reality" is synonymous with "objective reality".

Out of Einstein's dissatisfaction with the precipitate submission of the manuscript by Boris Podolsky (von Meyenn 2010, letter 206, p. 537), we conjecture that some flaws remained in the paper.

---

[4] Theoretical physics.

One more of these flaws is the sloppy use of the assignment of physical quantities to elements of physical reality, which we demonstrate by means of three statements from the paper.

Statement 1

... there is an element of physical reality corresponding to the physical quantity A.

Statement 2

The usual conclusion from this in quantum mechanics is that when the momentum of a particle is known, its coordinate has no physical reality.

Statement 3

In accordance with our criterion of reality, in the first case we must consider the quantity P as being an element of reality, in the second case the quantity Q is an element of reality.

There is a big difference in the nature of reality in the different statements above. If we apply the classification scheme for the structure of theories, we see that the latter statements would represent a naive realism where the referred physical quantity would be identical with the ontology of the theory.

The first statement that uses the correspondence between the physical quantity and the element of reality is more subtle. There, the physical quantity is a mathematical object with interpretation and corresponds to an independent ontological element of reality (see Table 4.1). The element of reality itself is only comprehensible through sensations, though it is expected to exist in the physical reality but never accessible to us in its whole nature.

It is coherent with Einstein's views on philosophy of science we presented above, and with the context of the EPR-paper, that he did not favour the point of view presented in statements 2 and 3.

4.2.3 Notions of Reality

Elements of Reality

In the EPR argument and the reality criterion specifically, a term is used to describe and speak about properties of objective reality: "element of reality." It is interesting to trace back the term "element" to a letter from Einstein to the philosopher Moritz Schlick in which he refers to Ernst Mach (Hentschel 1986):

Es scheint mir nun, daß das Wort 'wirklich' in verschiedenem Sinne genommen wird, je nachdem es von Empfindungen oder von Ereignissen bzw. Thatbeständen in physikalischem Sinne ausgesprochen wird. Wenn zwei verschiedene Völker unabhängig voneinander Physik treiben, werden sie Systeme schaffen, die bezüglich der Empfindungen ('Elemente' im Sinne Machs) gewiss übereinstimmen. Die gedankliche Konstruktion, die die beiden zur Verknüpfung dieser 'Elemente' ersonnen, können weitgehend verschieden sein.

It seems to me that the term "real"[5] *is used in different meaning, depending on whether used in physical context with sensations or on the other hand events respectively facts. When two nations pursue physics independently from each other, they will create systems that agree concerning the sensations(elements in the meaning of Mach). The construction of ideas both of them developed to connect these elements can differ widely.*[6]

Is this identification of the Machian elements mentioned by Einstein in the letter to Morith Schlick and the elements used in the EPR argument justified? In the Machian sense, an element is defined as sensation, as a representation for events (Mach 1922):

Die Ansicht, welche sich allmählich Bahn bricht, daß die Wissenschaft sich auf die übersichtliche Darstellung des Tatsächlichen zu beschränken habe, führt folgerichtig zur Ausscheidung aller müßigen, durch die Erfahrung nicht kontrollierbaren Annahmen, vor allem der metaphysischen (im Kantschen Sinne). Hält man diesen Gesichtspunkt in dem weitesten, das Physische und Psychische umfassenden Gebiete fest, so ergibt sich als erster und nächster Schritt die Auffassung der "Empfindungen" als gemeinsame "Elemente" aller möglichen physischen und psychischen Erlebnisse, die lediglich in der verschiedenen Art der Verbindung dieser Elemente, in deren Abhängigkeit voneinander bestehen.

" The view that wins recognition step by step, that science has to confine to clearly arranged representations of facts, leads to the exclusion of all dispensable, through experience not-controllable assumptions, specifically the metaphysical ones(in the meaning of Kant). Applying this view to physical and psychical fields, as first and next consequence results the view of sensations as common "elements" of all possible physical and psychical experiences. These sensations only exist in their different ways of connection between them and their mutual dependence of each other.[7]

and further

Berkeley[8] *sieht die "Elemente" als durch etwas außer denselben Liegendes, Unbekanntes (Gott) bedingt an, wofür Kant, um als nüchterner Realist zu erscheinen, das "Ding an sich" erfindet, während die hier vertretene Anschauung mit einer Abhängigkeit der "Elemente" von einander praktisch und theoretisch das Auskommen zu finden glaubt.*

" Berkley sees the "elements" as caused by something unknown(god) beyond them, Kant on the other hand, to appear as a down-to-earth realist, invents "the thing itself", while the here presented approach of mutual depending elements claims to get along for practical and theoretical purposes.[9]

---

[5] Refer to the discussion before on the terms "wirklich" und "real".

[6] Translation by the author.

[7] Translation by the author.

[8] Mach means the philosopher George Berkley.

[9] Translation by the author.

Machs philosophy of science is a positivistic one, an Empirio-criticism (Avenarius and Avenarius 1905; Avenarius 1906). Metaphysical parts of the theory are rejected as non-physical and unnecessary, also with regard to a principle of economy of thought all theories have to comply with. The elements Einstein is referring to are those parts of the theory that are indisputable because they represent the experiences, which are independent of the type of physical theory, in Einstein's relativistic context the invariants of a theory.

It is irony that Einstein refers to the originally positivistic assigned term "elements" to defend his realistic point of view, but he mentions also the Machian elements and gives a clear distinction between Machian elements and elements of reality (Einstein et al. 1935):

> *In a complete theory there is an element corresponding to each element of reality.*

The "element" in the theory is an element in the sense of Mach. The corresponding "element of reality" is not part of the theory; it is part of the ontology, part of the objective reality independent of our perception.

It is strange that the term "element" is never again used by Einstein; not in his explanation of the EPR argument in a letter to Schrödinger (von Meyenn 2010, Letter 206, p. 537), nor in the explanation in a letter to Popper (2005), nor in the reformulation of the EPR argument in Einstein (1948). This could give rise to the objection that the term "element" was coined by Podolsky who wrote most of the article as pointed out by Kiefer (2015, Chap. 2.4) and in von Meyenn (2010, Letter 206, p. 537).

The content of the EPR argument has been formulated after long discussions[10] of Einstein, Podolsky, and Rosen, so it seems feasible that the cornerstones of the paper were clearly defined before by all authors.

Secondly, though Einstein did not use the term in this context again, he mentions the "elements" before in the 1917 letter to Moritz Schlick (Hentschel 1986):

> *...a gap which widens progressively with the developing unification of the logical structure, that is with the reduction in the number of the logically independent conceptual elements required for the basis of the whole system.*

He further makes use of them in Einstein (1934):

> *It can scarcely be denied that the supreme goal of all theory is to make the irreducible basic elements as simple and as few as possible...*

It seems that Einstein had not agreed upon the way the EPR argument was constructed, paired with Podolsky's not-agreed submission of the EPR paper, and a parallel press statement raised Einstein's anger (Kiefer 2015). It is therefore plausible that the elements are indeed a reference by Einstein to the positivistic Machian "elements," amending them with ontological counterparts, the realistic "elements of reality," but Einstein did not want to refer to the EPR-paper and its terminology again.

---

[10] von Meyenn (2010, Letter 206, p. 537).

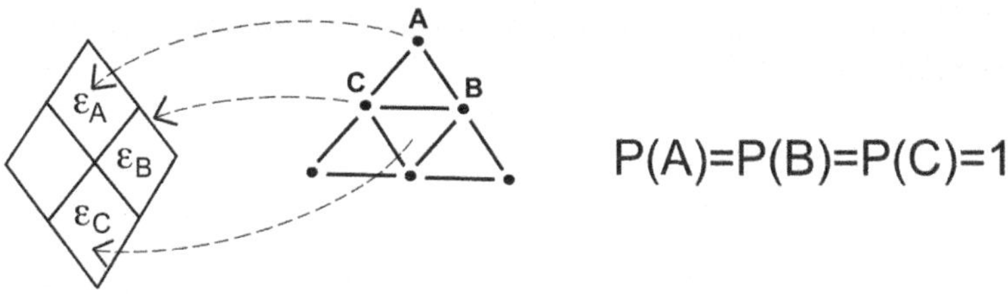

Fig. 4.3 Elements of reality

The Reality Criterion

The reality criterion is defined at the begin of the EPR paper to start the line of argument and identify physical quantities and concepts that have a counterpart in the objective reality. The problem is to identify those quantities and concepts that refer to reality and have a corresponding element in reality. For example, the concept of centre of mass of a body is not useful to describe some fundamental real property of the rigid body. It is merely a concept in the theory that is helpful to construct a plain theory and has some practical and technical advantages. So it is needed to find a way to identify elements in the theory that refer to an element in reality. Einstein admits that in his letter to Schrödinger (von Meyenn 2010, letter 206, p. 537):

> *Die eigentliche Schwierigkeit liegt darin, daß die Physik eine Art Metaphysik ist; Physik beschreibt "Wirklichkeit". Aber wir wissen nicht, was "Wirklichkeit" ist; wir kennen sie nur durch die physikalische Beschreibung!*[11]

> *The difficulty is due to that physics is a kind of metaphysics; physics describes "reality". But we do not know what "reality" is; we only know it through the physical description!*

The identified physical quantities that refer to the elements of reality are then chosen to show the incompleteness of quantum mechanics (Einstein et al. 1935) (Fig. 4.3):

> *If, without in any way disturbing a system, we can predict with certainty (i.e., with probability equal to unity) the value of a physical quantity, then there exists an element of physical reality corresponding to this physical quantity.*

We can define the reality criterion more formally by using the definitions from Eqs. 4.1 and 4.2.

One has to define a probability measure $P(t_i)$ for the elements of the theory. Then the reality criterion can be formally expressed as:

$$\forall t_i \in T_{PR} \land P(t_i) = 1 : \exists \epsilon_i \in PR \tag{4.3}$$

[11] Translation by the author.

By that we have a mapping ϕ that connects the element of the theory with the corresponding element of reality.

$$\phi: T_{PR} \rightarrow PR$$

$$\phi(t_A) = \epsilon_A$$

The remarkable thing in the way the reality criterion is defined is the working direction. Out of a probability statement in the theory, the reality criterion identifies corresponding elements of reality. If we assume classical physics that fulfills determinism, all physical quantities obey a probability[12] measure of one. Therefore, in classical physics, all physical quantities should have an corresponding element of reality. The situation is not that trivial, as an example will show.

Let us assume Newtonian mechanics in one dimension to describe the motion; i.e. that is, the position of a particle A in reference to a second particle B. The particle is assumed to have a constant mass and is described by a mass point. Our theory then contains the following elements: Occurring forces, the mass of the particle, the position coordinate of particle A and the position coordinate of particle B.

$$T_{PR} = \{F, \ m, \ x_A(t), \ x_B(t)\} \tag{4.4}$$

The theory will con nect these elements with Newton's second law to describe the motion of particle A:

$$F = m\frac{d^2x_A(t)}{dt} \tag{4.5}$$

Let us assume the simplest situation; the force vanishes and becomes zero. This differential equation delivers the equation of motion with two independent constants that refer to the initial conditions. They turn out to be the initial speed of particle A and the initial position of particle A.

$$x_A(t) = v_{Ao} \cdot t + x_{Ao} \tag{4.6}$$

Our theory is deterministic; therefore, the probability measure is equal to one, so by applying the criterion of reality we can map all elements of the theory to elements of reality with a mapping ϕ. That way we have identified corresponding elements of reality $PR = \{\epsilon_F, \epsilon_m, \epsilon_{x_A(t)}, \epsilon_{x_B(t)}\}$ (Fig. 4.4).

But there is a problem. Coordinates can be transformed to a different reference frame. Let us apply Galilean transformations on the elements of our theory.

$$x' = x(t) + v \cdot t + a \tag{4.7}$$

[12] We neglect all interpretational issues concerning probability here and consider probability in classical physics as a propensity.

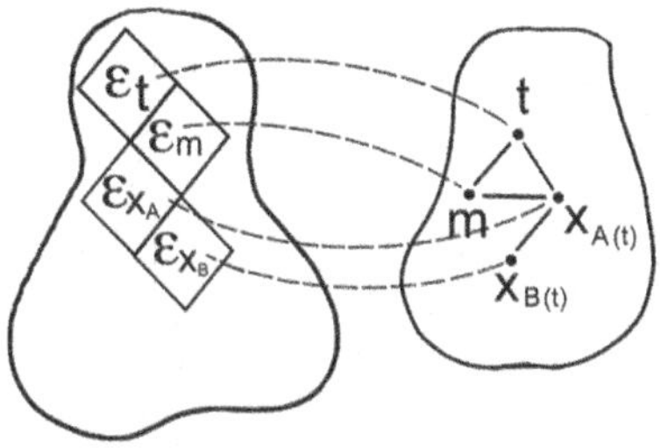

Fig. 4.4 Newton's theory mapped to elements of reality

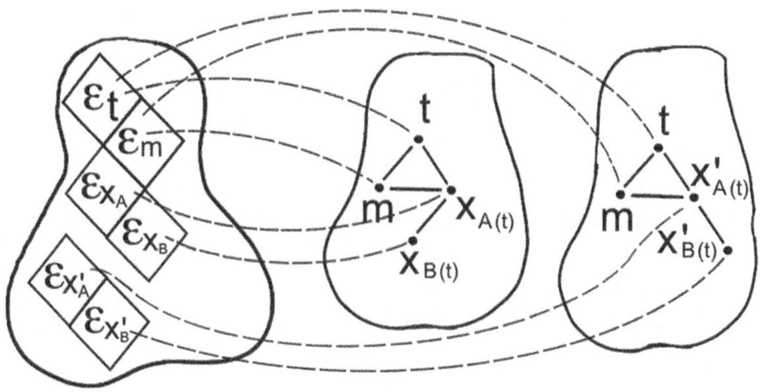

Fig. 4.5 Newton's theory mapped to elements of reality

Without going into details, it turns out that force and mass will be invariant under Galilean transformation, but we receive transformed elements of the theory for $x_A(t)$ and $x_B(t)$. A mapping of the transformed theory T' delivers new elements of reality (Fig. 4.5).

This mapping shows us that only by using deterministic elements of the theory, and therefore fulfilling Einstein's criterion of reality, we are not able to identify elements of reality that are really satisfying, because they cannot be identified as unambiguous.

What can be done to resolve this unsatisfying situation is to choose the elements of the theory differently. The position of the particle A can also be described by the distance to the particle B.

$$d(t) = x_A(t) - x_B(t) \tag{4.8}$$

It turns out that this distance $d(t)$ is invariant under Galilean transformation and the mapping ϕ and ϕ' of the theory T and its Galilean-transformed alternative T'. We therefore obtain elements of reality that are satisfying due to their invariance (Fig. 4.6).

The chosen example was simple to demonstrate this idea, but can be transferred to the theory of relativity or any other classical theory. Especially in the theory of

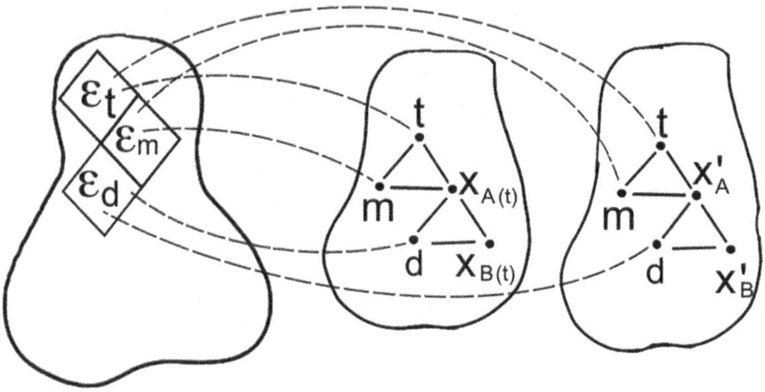

Fig. 4.6 Newton's theory mapped to elements of reality with invariants

relativity, the concept of invariance (covariance) will play a foundational role in the conception of the theory. Einstein refers to this invariance principle when he states (Einstein 1936):

> *Nur solche Gleichungen sind als Ausdruck allgemeiner Naturgesetze sinnvoll, welche bezüglich in diesem Sinne beliebiger Transformationen kovariant sind. (Postulat der allgemeinen Kovarianz).*
>
> *Only such equations that fulfil covariance regarding any transformation are resaonable to represent laws of nature. (Postulate of general covariance).*[13]

The reality criterion is a method to find those invariant elements in the theory to identify corresponding elements of reality that have ontological character.

Lehner emphasizes the role of invariants in Lehner (2014, p. 313):

> *The association of descriptive invariants with objective reality has become a methodological standard in modern physics: it is the conceptual basis not only of general relativity, but of all forms of gauge theories, also in their quantum forms.*

This key role of invariances will encourage Lehner to define a methodological realism he ascribes to Einstein's philosophy of physics, which we discuss in Sect. 4.2.3.

Another aspect of the reality criterion is worth a discussion. Fine (2009) states that the reality criterion can only be found in the EPR paper and was not used by Einstein in any other presentation of the EPR argument:

> *...one should note that although Einstein himself later published several versions of EPR, none of them makes any reference to or use of that reality criterion.*

We can show that this is not correct. In his letter to Karl Popper, Einstein presented the EPR argument in a compact way, but clearly made use of the reality criterion when he states Popper (2005, p. 481):

---

[13] Translation by the author.

> *For if, upon freely choosing to do so [that is, without interfering with it], I am able to predict something, then this something must exist in reality.*[14]

The reality criterion is relevant for one main agenda in the EPR paper; to criticise the statistical character of quantum mechanics.

The Statistical Character of Quantum Mechanics

In the EPR paper, no direct reference can be found to answer the question if in the EPR argument the quantum-mechanical description refers to a statistical ensemble or to single events, to a single quantum-mechanical object. It is not discussed explicitly, though one can conclude for good reason that a single object is meant when they repeatedly speak about a particle's behaviour (Einstein et al. 1935, p. 779):

> *The fundamental concept of the theory is the concept of state, which is supposed to be completely characterized by the wave function ψ, which is a function of the variables chosen to describe the particle's behaviour.*
>
> *For this purpose let us suppose that we have two systems, I and II,... We can then calculate with the help of Schrödinger's equation the state of the combined system I + II at any subsequent time;*

Using the term "system," the authors left open whether they apply the quantum-mechanical description to an statistical ensemble or to a single quantum-mechanical object. One year later, after several discussions and after a reply paper by Bohr, Einstein had chosen a clear position on this question (Einstein 1936):

> *Es scheint mir deshalb klar, dass die Born 'sche statistische Deutung der Aussagen der Quantentheorie die einzig mögliche ist: Die ψ-Funktion beschreibt überhaupt nicht einen Zustand, der einem einzelnen System zukommen könnte; sie bezieht sich vielmehr auf so viele Systeme, eine "System-Gesamtheit" im Sinne der statistischen Mechanik.*
>
> *It is obvious to me that the statistical interpretation of quantum theory by Born is the only way: The ψ-function does not describe a state that represents one single system; It refers to many systems, to a system-totality in the sense of statistical mechanics.*[15]

Even more, he refers to the EPR paper and states that the ensemble interpretation solves any difficulty (Einstein 1936):

---

[14] The original wording in German is slightly different and we would prefer to give a different translation:

> *Denn was ich nach freier Wahl prophezeien kann, das muss auch in der Wirklichkeit existieren.*
>
> *What I can prophesy by free choice must also exist in reality*

[15] Translation by the author

> *Eine solche Interpretation beseitigt auch eine von mir zusammen mit zwei Mitarbeitern jüngst dargestellte Paradoxie ... Die Zuordnung der ψ-Funktion zu einer System-Gesamtheit beseitigt auch hier jede Schwierigkeit.*
>
> *Such an interpretation also removes a recently by me and two colleagues presented paradox ... The assignment of the ψ-function to a system-totality here also eliminates all difficulties.*[16]

It is not excactly defined what Einstein had in mind with these "difficulties", but it is most likely that he refers to critiques on the EPR paper Bohr (1935), (Ruark (1935). Without any doubt the difficulties do not refer to the question if quantum mechanics is complete or incomplete. That position of Einstein is unchanged as one can see from the presented statement concerning the EPR argument one year later (Einstein 1936):

> *Diese Zuordnung mehrerer ψ-Funktionen zu demselben physikalischen Zustande des Systems B zeigt wieder, dass die ψ-Funktion nicht als (vollständige) Beschreibung eines physikalischen Zustandes (eines Eingelsystems) gedeutet werden kann.*
>
> *This assignment of several ψ-functions to one physical state of the system B demonstrates that the ψ-function cannot be interpreted as a (complete)description of a physical state (of a single quantum mechanical object)*[17]

Einstein changed his position on the statistical character on quantum mechanics due to critique. Of course, he was aware of the statistical character of quantum mechanics before,[18] but the Born interpretation was an argument for incompleteness that was much easier to defend than the EPR argument.

The statistical character of quantum mechanics is connected to the notion of completeness. We see in Sect. 4.3.3 that there are several definitions of completeness and there is a deep connection to the statistical character of quantum mechanics. Based to the previously presented aspects we analyse cornerstones of Einstein's realism.

---

[16] Translation by the author.

[17] Translation by the author.

[18] In his letter to Schrödinger where he explained the EPR argument, he also refers to the statistical character (von Meyenn 2010, letter 206, p. 537):

> *Nun beschreibe ich einen Zustand so: Die Wahrscheinlichkeit dafür, daß die Kugel in der ersten Schachtel ist, ist $\frac{1}{2}$ Ist dies eine vollständige Beschreibung? Nein: Eine vollständige Aussage ist: die Kugel ist in der ersten Schachtel (oder ist nicht). So muß also die Charakterisierung des Zustandes bei vollständiger Beschreibung aussehen.*
>
> *I describe a state this way: The probability for a ball being in the first box is $\frac{1}{2}$. Is that a complete description? No: A complete description is: The ball is in the first box (or is not). The representation of a complete state has to be that way." Translation by the author.*

Cornerstones of Einstein's Realism

The notion of reality cannot be seen isolated; it is connected to several conceptions that accompany reality. Neither can a conception of reality by a scientist be seen as unvarying; it is subject to change and variation and often inconsistent.[19] A thorough discussion of Einstein's realism would go beyond our scope; therefore, we only present aspects or cornerstones of the ideas of reality that Einstein had in mind. For a more detailed analysis of the evolution of Einstein's philosophical position over the years, refer to Howard (1993), an in-depth discussion on Einstein's view on realism and philosophy of science is given by Janssen and Lehner (2014).

- Objectivity—There is a world outside, independent of our experiences, descriptions and ideas, a given reality that awaits to be uncovered. Connected to that, Einstein also has in mind that there is a description that reflects this truth about nature. In that sense, this view is contradictory in its claim to represent elements of the theory as human thoughts and ideals and simultaneously follow a correct path, as Einstein calls it. See Sect. 4.1 "Einstein's view on theories and theory formation".
- (Mathematical)—Simplicity: Einstein has in mind that a physical theory has to fulfill simplicity (Einstein 1934, p. 165):

  > *It can scarcely be denied that the supreme goal of all theory is to make the irreducible basic elements*[20] *as simple and as few as possible without having to surrender the adequate representation of a single datum of experience.*

  Moreover, he assumed this simplicity is something that is realised in nature as well (Einstein 1934, p. 165):

  > *Our experience up to date justifies us in feeling sure that in Nature is actualized the ideal of mathematical simplicity.*

  This is a contradiction in the understanding of simplicity, because the ideas and elements of the theory should fulfill simplicity, but they are merely a construction by human minds, so this simplicity is nothing fundamental and cannot be the same simplicity he mentions in the second quotation. We present a discussion of the concept of simplicity in context with interpretations of quantum mechanics and its peculiarities in Sect. 5.4. For an account on Einstein's conception of simplicity, refer to Howard (1993, p. 230).
- Mathematical entities—Einstein favours the four dimensional continuum over algebraic structures with arbitrary or even infinite dimensionality. Out of some statements, one could conclude that he took the position of naive realism in a sense of taking mathematical entities as elements of reality, a mathematical realism.

---

[19] By this inconsistency we mean the contradiction between several ontological statements or concepts that refer to reality.

[20] By elements he refers to the Machian elements of the theory.

- Methodological and phenomenal realism—According to Lehner (2014), Einstein advanced a view on the role of invariants in theoretical models; he calls that position methodological realism. The role of invariants in a theory is a distinguished one; the invariants represent elements of the theory that are assumed to have a strong connection to objective reality. This conception is based on the theory of relativity, but also found application in modern theoretical physics like in Gauge theories. In contrast to methodological realism, he sees phenomenal realism (Lehner 2014, p. 313):

  > *By contrast, physical measurements do not reveal those objective entities themselves, but their representations in local coordinate frames, that is, non-invariant quantities. Such measurements are no less real, but they are real in a different sense: their reality is phenomenal, dependent on an observer situated in a specific way.*

  Following Lerner's reading, it was Einstein's attempt to apply methodological realism to quantum mechanics, and the failure of this attempt was reason for rejecting quantum mechanics as a foundational theory.
- Deterministic character—Einstein was very clear about that point. To him, a theory with statistical character is only provisional and cannot represent real events. Refer also to the discussion in Sect. 4.2.3.
- Non-localizability: In some sense, this non-localizability is a nonlocality feature, but not in the sense of an action-at-a-distance; more in the sense of the notion of a non-absolute localization of particles. The concept of non-localization was familiar to Einstein from the way the gravitational energy behaves in general theory of relativity. We provide a more detailed discussion of Non-localizability in Sect. 4.3.2.
- Singularity-free—To Einstein the representation of particles in a field theory implied the problem of singularities of the field. To him a reasonable representation of particles has to be free from such singularities in the field. Refer also to the discussion in Sect. 4.3.2.
- Separability: The physical reality allows separation of the world into parts that can be handled independently. To Einstein, this is a prerequisite for scientific reasoning per se (See Don Howard Howard 1993, p. 234ff; Howard 2005, p. 39). The spatiotemporal separability consists of locality and separability of systems, per se. There is also a separability without a spatiotemporal aspect, which is realised, for instance, by linear systems that can be decomposed into systems that evolve independently of each other, even if they share a position in spacetime. Fourier analysis is a good example of this superposition principle that allows decomposition of systems into separable subsystems.
- Locality and causality—In Einstein's view, reality obeys locality in the sense of local realism: Events in two spacelike separated areas cannot affect each other. Therefore, all meaningful descriptions of nature have to obey local realism as well.

- Unambiguity or uniqueness: A complete description[21] leaves no room for ambiguity. This is connected to Aristotelian ideas of physics—basically the contradiction principle—which is based on common sense experience. This view reflects also in the EPR argument, in the condition of completeness (Einstein et al. 1935) and in later variations of the EPR argument (von Meyenn 2010, p. 537; Popper 2005, p. 482; Einstein 1936, p.341; Einstein 1948).

4.3 EPR=ER in 1935 and Its Revival

In this section, we provide an answer to the question of whether there is a connection between the EPR and Einstein-Rosen (ER) paper and, if yes, in what kind of way. In a modern context, where in some interpretations of quantum mechanics nonlocality plays a crucial role, and the imprint by popular culture, it is a tempting thought to put the nonlocality features of quantum mechanics in connection to nonlocal structures like wormholes. It is even more surprising to find out that the first wormhole paper[22] was written by the same persons that challenged the completeness of quantum mechanics with the EPR argument, namely Einstein and Rosen. The conclusion one could draw at first glance is close-at-hand: Einstein and Rosen could have had in mind to solving the issue with nonlocalities in quantum mechanics by wormholes. This conclusion is unjustified due to several reasons we present now.

4.3.1 Nonlocality

At first, one has to say that there was no issue with nonlocality to solve; at least, not for Einstein. Einstein recognised the nonlocality of quantum mechanics, which basically occurred in context with the projection postulate, but to him it was untenable to accept nonlocalities in physics, so he was not searching for a way to explain nonlocalities by an underlying theory. It is this following statement that makes clear that nonlocality, for the authors of the EPR-paper, and it can be assumed especially for Einstein, was not considered to be a possible or even thinkable part of ontology (Einstein et al. 1935):

> *This makes the reality of P and Q depend upon the process of measurement carried out on the first system, which does, not disturb the second system in any way. No reasonable definition of reality could be expected to permit this.*

---

[21] See Sect. 4.3.3 for an overview on several definitions of completeness. This completeness will be defined as *theory completeness* and will refer to the elements of the theory.

[22] Before the ER paper, only speculations by Flamm and Weyl can be found. As pointed out by Visser (1996) who coined the term "first wormhole paper".

The action-at-a-distance effects in Newtonian mechanics were already a problem to Newton and entail a lot of philosophical problems, such as difficulties with causality. It is comprehensible that Einstein was not considering to give up the beautiful new and self-consistent formulation of a causal structure. To him, there must have been a different solution that would integrate the quantum effects in the framework of field theory with its complete[23] character. But field theory had its problems as well, and Einstein was full aware of them.

If nonlocality was not the pursued goal of the ER-paper, what was it then? In his paper on the method of theoretical physics, he clarifies one of the main difficulties in field theory (Einstein 1934):

> *The most difficult point for such a field-theory at present is how to include the atomic structure of matter and energy. For the theory in its basic principles is not an atomic one in so far as it operates exclusively with continuous functions of space, in contrast to classical mechanics whose most important feature, the material point, squares with the atomistic structure of matter.*

The absolute localization of particles was an open question to Einstein, and we demonstrate the connection to the ER-paper in Sect. 4.3.2.

4.3.2 Localizability of Particles

In Sect. 4.1, we presented Einstein's understanding of the term "atomic structure": The singularity-free representation of particles in space. To him, it was not only a problem of field theory, but had a deep connection to quantum theory (Einstein 1934):

> *... it seems to me certain that we have to give up the notion of an absolute localization of the particles in a theoretical model. This seems to me to be the correct theoretical interpretation of Heisenberg's indeterminacy relation.*

Not only that the particle is represented singularity-free, the refutation of an absolute localization of the particle was the goal of this concept.[24] To Einstein, it promised the reconstruction of quantum effects by general theory of relativity.

The ER-paper opens in the abstract with reference to this atomistic theory of matter (Einstein and Rosen 1935b):

> *The writers investigate the possibility of an atomistic theory of matter and electricity which, while excluding singularities of the field, makes use of no other variables than the $g_{\mu\nu}$ of the general relativity theory and the $\phi_{\mu\nu}$ of the Maxwell theory.*

In the model of the Einstein-Rosen bridge, the bridge is a representation of a particle, but the position of the particle is smeared out. The position is in contrast to the external

[23] In the sense of *theory completeness.*

[24] It is noticeable that this highly non-classical idea has been proposed by Einstein, who repeatedly had to face critique on his inflexible position and his adherence to classical physics.

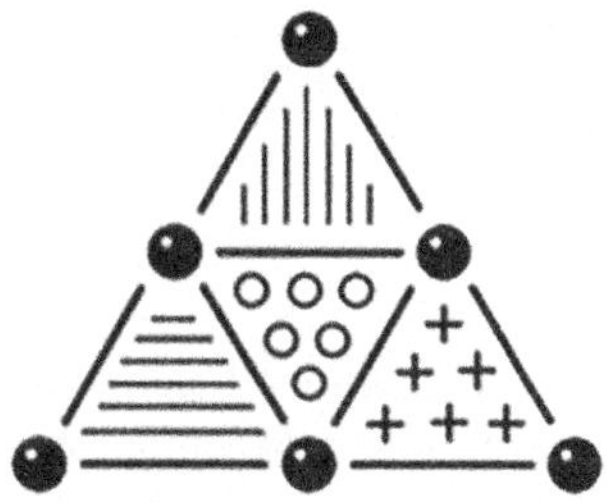

Fig. 4.7 Structure of a theory

Schwarzschild-solution not given by the position coordinate of the singularity. The metric around the particle is defined by the Einstein-Rosen-bridge solution; still, there is no account of the particles exact position. This is satisfying in the sense of Einstein's requirement for solving the quantum problem within a continuum theory presented in Sect. 4.1 (Einstein 1934).

It is comprehensible that the goal of the ER-paper was to introduce a field theory program that solves the localisation problem for particles. Even more, is was pursued to reconstruct quantum theory with a unified field theory that can claim to be complete. The definition of completeness and its peculiarities will be presented in Sect. 4.3.3.

For a more detailed discussion concerning the philosophical implications of the concept of particles, refer to Achinstein (1991), Kuhlmann (2010).

4.3.3 Completeness

The notion of completeness used in the EPR argument, ER paper and accompanying sources is a peculiar concept. While initially seeming straightforward, a precise definition is essential for contextualizing the EPR argument. We will demonstrate that multiple definitions of completeness are at play and elucidate their connection to the ER paper.

Theory Completeness

In 1934, Einstein defined the meaning of the term completeness (Einstein 1934):

> *A complete system of theoretical physics consists of concepts and basic laws to interrelate those concepts and of consequences to be derived by logical deduction.*

In Sect. 2 a notion of the structure of a physical theory has been presented. By using physical quantities, laws to connect them, and concepts to give structure and principles to the theory, a complete theory in this sense can be pictured as following (Fig. 4.7):

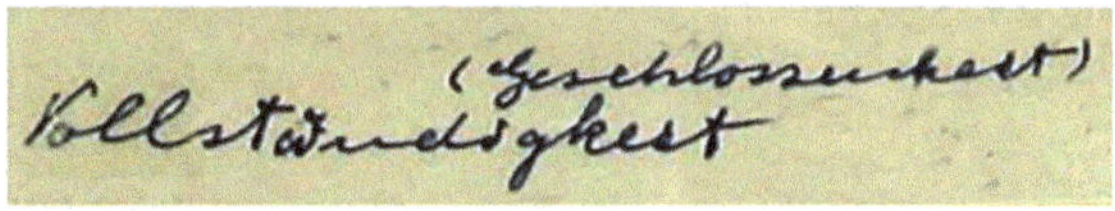

Fig. 4.8 Completeness—Closeness (Einstein and Rosen 1935a)

The mentioned elements (physical quantities, laws and concepts/principles) are meant in the Machian sense and not connected, per se, to an ontology. Einstein emphasized that these elements are "...free inventions of the human mind..." (Einstein 1934) and therefore pure constructions of thought. Therefore, this definition of completeness can be understood more as a consistency in the theoretical system. In any sense, it forgoes to provide ontological statements. We call this completeness *theory completeness*.

An application of this theory completeness is given in Einstein and Rosen (1935b):

> *One of the imperfections of the original relativistic theory of gravitation was that as a field theory it was not complete; it introduced the independent postulate that the law of motion of a particle is given by the equation of the geodesic. A complete field theory knows only fields and not the concepts of particle and motion. For these must not exist independently of the field but are to be treated as part of it.*

We want to give a definition for theory completeness that gets as close as possible to the concept Einstein had in mind:

Definition 4.1 A theory fulfills theory completeness if its physical quantities and laws are connected by concepts that relate only to each other, and the theory renounces extra assumptions of concepts that exceed the framework of the theory.

Theory completeness does not refer to an ontology. It only refers to elements of the theory, and requests a consistency in the used concepts. Einstein refers to it in Einstein and Rosen (1935b) as Closeness (Vollständigkeit) (Fig. 4.8).

References to *theory completeness* can be found in: Einstein (1934, p. 164, 166), Einstein et al. (1935, p. 778), Einstein and Rosen (1935b, p. 76, p. 77), Einstein (1936, p. 316, p. 344), Einstein (1948, p. 320, p. 323).

Bijective Completeness

The EPR paper posed a question on completeness, but is it *theory completeness* that is challenged? If we refer to the EPR paper, a definition of completeness is given Einstein et al. (1935) (Fig. 4.9):

> *Whatever the meaning assigned to the term complete, the following requirement for a complete theory seems to be a necessary one: every element of the physical reality must have a counterpart in the physical theory*

As can be seen immediately, this definition of completeness differs from the definition of *theory completeness*; it even further extends the term completeness to the ontological level, whereas the *theory completeness* definition refers only to the theory

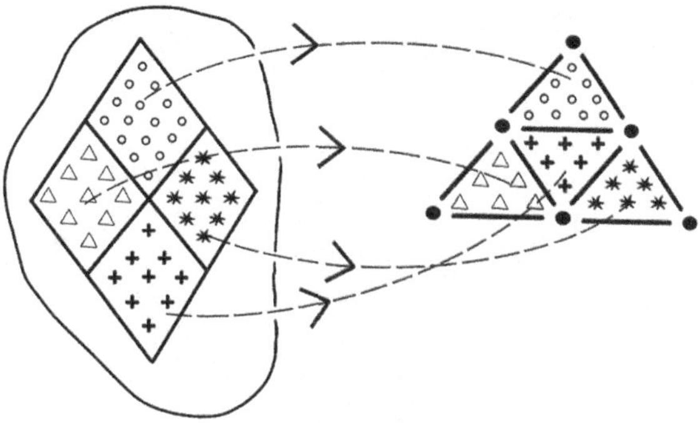

Fig. 4.9 Bijective completeness

and its elements. The latter relates the elements of reality to the elements of a theory. Fine (2009) calls this completeness *bijective completeness*. Following Einstein et al. (1935), a definition would be:

Definition 4.2 Bijective completeness means that every element of the physical reality ϵ_i must have a counterpart t_i in the physical theory.

Though *bijective completeness* is a Bijection between elements of the theory and elements of reality, it starts with the latter ones. Starting at elements of reality causes the problem how of to identify them and the corresponding elements of the theory. For this purpose, Einstein et al. (1935) had to assume a criterion of reality to identify these elements of reality.[25]

From an accompanying source, one more aspect about *bijective completeness* can be seen. Karl Popper received a letter from Einstein in September of 1935, where Einstein explained the idea of the EPR paper in a short and very clear way. One more aspect of Einstein's notion of *bijective completeness* can be found there (Popper 2005, p. 413):

> *Since a complete description of a physical state must necessarily be an unambiguous description (apart from superficialities such as units, choice of the co-ordinates etc.), it is therefore not possible to regard the ψ-function as the complete description of the state of the system.*

He concludes that a *bijective completeness* description has to be unambiguous, so if every element of the physical reality has to have a (one) correspondence in the theory, it must be vice versa. This unambiguousness assumption is not presented explicitly as with the reality criterion, the completeness criterion, or separability/locality, though it plays an important role in the EPR argument (Fig. 4.10).

It is further interesting that Einstein seems to have seen a problem with the pure *bijective completeness* argument in the EPR paper. Superficialities as choice of coordinates, as we have presented them before in Sect. 4.2.3 would not fulfill *bijective completeness*, so Einstein clarifies this to Popper (2005, p. 413).

Einstein gives no exact definition of completeness in the EPR paper, but states that Definition 4.2 of *bijective completeness* is a necessary requirement for his concept

---

[25] See also Sect. 4.2.3.

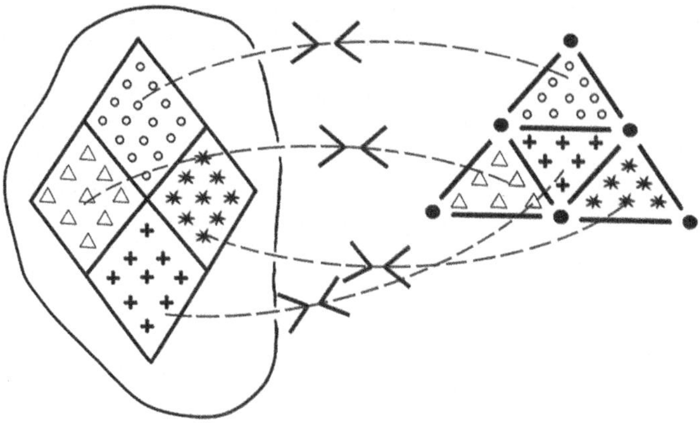

Fig. 4.10 Bijective completeness and unambiguousness

of completeness. Is it possible that *theory completeness* is the completeness Einstein had in mind? *Theory completeness* is only defined by elements of the theory, and makes no statement about the ontology of the theory; therefore it can be ruled out that *bijective completeness* is a necessary requirement of *theory completeness*. *Bijective completeness* would have to be true, so that *theory completeness* could be true, but it is obvious that the structure of the elements of a theory is completely independent of any mapping ϕ between the elements of the theory and elements of reality. *Theory completeness* is not the completeness Einstein had in mind, for which he used *bijective completeness* as a necessary requirement.

References to the *bijective completeness* can be found in Einstein et al. (1935, p. 777, p. 778, p. 780), Einstein (1936, p. 341).

Born Completeness

Since *theory completeness* is excluded from being the completeness meant in the EPR paper, there must be a different definition of completeness. Fine (2009) identifies two further concepts of completeness: *Born completeness* and *Schrödinger completeness*:

> *According to the Bornian conception, a complete description is essentially nonprobabilistic; genuinely probabilistic assertions are necessarily incomplete.*

These conceptions of completeness can be identified in letters from Einstein to Schrödinger, where Einstein also stated the "Einstein's boxes" Gedankenexperiment, which discusses the probabilities of finding a ball in a box (von Meyenn 2010, letter 206, p. 537). Kupczynski (2006) introduces the notion of *predictable completeness* to describe this relation. A theory might predict all phenomena observed in experiments correctly without yet grasping their fine structure. Such a theory would not be considered predictable complete and a re-evaluation of the experimental data would lead to the discovery of additional variables governing this fine structure. Both *Born completeness* and *Schrödinger completeness* imply *predictable completeness.*

Definition 4.3 Born completeness for a theory is fulfilled if the theory contains no probabilistic elements. A theory which contains statistical accounts and probabilistic statements is considered to be incomplete.

Following this definition, quantum mechanics is incomplete per definition. No further arguments are needed.

Is *Born completeness* the completeness Einstein meant in the EPR paper? If *Born completeness* would be addressed, it would mean that the authors of the EPR paper had in mind that without *bijective completeness*, *Born completeness* could not be true. That follows from the EPR condition of completeness.

On the other hand, the statement "quantum mechanics fulfills *Born completeness*" is false. In this case, out of the rules for the material conditional, it follows that independent of the truth value of the *bijective completeness*, the statement "*theory completeness* is a necessary requirement for *Born completeness*" is true. To show the truth value of *bijective completeness* is therefore meaningless, and we can rule out that *Born completeness* is meant as the conception of completeness in the EPR paper.

References to the *Born completeness* can be found in Einstein (1936, p. 339, p. 341, p. 343, p. 347).

Schrödinger Completeness

Fine (2009, p. 71) gives an analysis of *Schrödinger completeness*:

> *By contrast the Schrödinger view is that probabilities can be fundamental, not to be reduced to something else. Thus the Schrödinger conception is that a complete description of a state of affairs can be a probabilistic assertion, with probability less than unity, which (somehow) tells the whole truth about that state of affairs. If there were some further truth to be told, then the probabilistic assertion would be an incomplete description*

It's getting a bit vague with *Schrödinger completeness* because it seems that Einstein's conception of it is not a very clear one; at least, he did not define it explicitly. What can be seen immediately is that the definitions of probability in context with *Born completeness* and *Schrödinger completeness* differ vastly. Whereas the concept of the probability of occurrence of outcomes in an ensemble of events is used in the first conception, the latter defines probability as fundamental and related to single events. The interpretation of probability will go beyond the scope of this work. We therefore stick to an operational definition of *Schrödinger completeness* that is based on Einstein's considerations in his letter to Schrödinger (von Meyenn 2010, letter 206, p. 537):

Definition 4.4 A theory containing probabilistic statements about elements of the theory is Schrödinger complete if those statements refer to single events or entities and are fundamental in the sense that nothing more can be said about the system.

The theory is considered to be incomplete if there are potential new elements[26] in the theory that would abandon the probabilistic character of the theory.

The definition of *Schrödinger completeness* refers to what we would nowadays call "hidden variables."

It is difficult to decide if *Schrödinger completeness* is the completeness Einstein is referring to in the EPR paper. If we carefully analyse the argument in the letter to Schrödinger from the 19th of June 1935, we find a self-reference in the definition of completeness (von Meyenn 2010, letter206, p.537):

> *Nun beschreibe ich einen Zustand so: Die Wahrscheinlichkeit dafür, daß die Kugel in der ersten Schachtel ist, ist $\frac{1}{2}$ Ist dies eine vollständige Beschreibung? Nein: ... Ja: ... Der Zustand vor dem Aufklappen ist durch die Zahl $\frac{1}{2}$ vollständig charakterisiert, deren Sinn sich bei Vornahme von Beobachtungen allerdings nur als statistischer Befund manifestiert.*
>
> *I describe a state this way: The probability for a ball being in the first box is $\frac{1}{2}$. Is that a complete description? No: ... Yes: ... The state before opening the box is completely characterised by the number $\frac{1}{2}$, whose meaning becomes manifest as statistical account by applying measurements on the system.*

What is meant by this complete characterisation by the number $\frac{1}{2}$? If we assume it to be *bijective completeness* in a way that the number is ascribed to an element of reality, and the number is interpreted as probability in an epistemic way, this would represent an epistemic account of the situation. Depending on the interpretation of probability, it would represent propensity, knowledge about the state of the ball, or degrees of belief on its state.

In principle, this would be a description that would fulfill *bijective completeness*, because one can give a Bijection to elements of reality in each of these accounts. Therefore, it is plausible to identify *Schrödinger completeness* as the completeness Einstein challenged in the EPR paper.

ψ—Completeness

In an account to classify hidden variable models, Harrigan and Spekkens (2010, p. 131) proposed a definition of *ψ-completeness*:

> *An ontological model is ψ-complete if the ontic state space Λ is isomorphic to the projective Hilbert space $\mathcal{PH}$ (the space of rays of Hilbert space) and if every preparation procedure P_ψ associated in quantum theory with a given ray ψ is associated in the ontological model with a Dirac delta function centered at the ontic state Λ_Ψ that is isomorphic to Ψ, $p(\lambda|P_\psi) = \delta(\lambda - \lambda_\psi)$.*

---

[26] Einstein calls them "fremde Faktoren" (external factors). The translation of "fremd" is peculiar, since in the literal translation it would be "strange", "weird", "alien" or "different." But in context, it refers to factors that are not considered in the theory. They are not part of the theory, so we decided to use the non-literal translation "external." In our modern context, these factors could be understood as "hidden variables."

This definition of *ψ-completeness* is claimed to be identical with *bijective completeness* (Harrigan and Spekkens 2010, p. 147):

> *It is quite clear that by "real state of the real system", Einstein is referring to the ontic state pertaining to a system. Bearing this in mind, his definition of completeness can be identified as precisely our notion of ψ-completeness given in Definition 2*

This is a misunderstanding of Einstein's view on completeness and elements of reality. Einstein is not using the wording "elements of reality" in his letter to Schrödinger (von Meyenn 2010, letter206, p. 538), but referring to *bijective completeness*:

> *Man beschreibt in der Quantentheorie einen wirklichen Zustand eines Systems durch eine normierte Funktion ψ der Koordinaten (des Konfigurationsraumes).Die zeitliche Änderung ist durch die Schrödinger-Gleichung eindeutig gegeben. Man möchte nun gerne folgendes sagen: ψ ist dem wirklichen Zustand des wirklichen Systems eindeutig zugeordnet. Der statistische Charakter der Meßergebnisse fällt ausschließlich auf das Konto der Messapparate bzw. des Prozesses der Messung. Wenn dies geht rede ich von einer vollständigen Beschreibung der Wirklichkeit durch die Theorie. Wenn aber eine solche Interpretation nicht durchführbar ist, nenne ich die theoretische Beschreibung *unvollständig".*

> *In quantum theory a real state of a system is described by a normed function ψ of coordinates (of configuration space). The evolution in time is given by Schrödinger equation uniquely. One would like to say: ψ is corresponding to the real state of the real system uniquely. The statistical character of the measurement results is a consequence of the measurement apparatus respectively the measurement process. If this works, I speak about a complete description of reality by a theory. If such an interpretation is not possible, I call the theoretical description incomplete.*[27]

Harrigan and Spekkens (2010) conclude out of these statements that Einstein's *bijective completeness* is identical to their *psi-completeness*, but the latter one is setting up an Isomorphism between the projective Hilbert space and an ontic state space, whereas *bijective completeness* is a Bijection between elements of a theory and elements of reality. These elements of reality have no defined mathematical structure, per se. In our proposed classification scheme for interpretations of quantum mechanics, the elements of reality are located in level four, the ontological level. Assuming that the elements of reality are elements of an ontic state space, is a kind of mathematical realism, but we think that is not the intention of the claim.

On the other hand, Harrigan and Spekkens (2010) have good reasons to assume that Einstein, by referring to the "real state of the real system" (von Meyenn 2010, letter 206, p. 538), means elements of a theory that would replace quantum mechanics. They propose an argument for Einstein's preference for the EPR argument, in contrast to his incompleteness argument from 1927. One aim, according to them, was Einstein's goal to emphasize the epistemic character of the state vector and to look out for hidden variable theories.

ψ-completeness is a useful classification tool for hidden variable theories, but we refuse the identity of *ψ-completeness* with *bijective completeness*, based on the involvement of completely different structures in these two conceptions. This does not affect any applications of *ψ-completeness* in Harrigan and Spekkens (2010).

---

[27] Translation by the author.

Standard Completeness

The following completeness criterion is not part of the original EPR discussion, and is presented here to amend the overview on accounts on completeness. Held (2012) gives an argument against *standard completeness*. He brings the timeframe of the involved events into consideration and claim a conflict between *standard completeness* and the principle that "QM should deliver probabilities for physical systems possessing properties at definite times" (Held 2015).

Remarks on Completeness and the EPR-ER Connection

The targets of the EPR paper are manifold; one aim is to show that *Schrödinger completeness* does not hold for the unitary core of quantum mechanics itself. Einstein claimed that quantum mechanics did not receive its statistical character out of the measurement process. The unitary evolution of the state vector itself is, according to him, incomplete in the sense of *Schrödinger completeness*. The critique on quantum mechanics is one story, but more interesting in our context is the motivation for showing this incompleteness. Replacing a successful theory with a new unknown theory without a good reason encounters opposition. Einstein had to convince the community of the incompleteness of quantum mechanics to present a new program (Einstein et al. 1935):

> *While we have thus shown that the wave function does not provide a complete description of the physical reality, we left open the question of whether or not such a description exists. We believe, however, that such a theory is possible.*

Einstein retreated quickly, already in 1936, to the position of *Born completeness*, which to him was sufficient to show the need for a theory that fulfills his requirements. Referring to his article Einstein (1948), he notes in a letter to Besso (Speziali et al. 1972, p. 403):

> *Es freut mich dass Du meinen kleinen Aufsatz gelesen hast. Hast Du auch gemerkt wie unlogisch Pauli darauf geantwortet hat? Er leugnet es, dass diese Art der Beschreibung unvollständig sei, sagt aber im selben Atemzuge, dass die ψ Funktion eine statistische Beschreibung des Systems sei, die Beschreibung einer System-Gesamtheit. Dies ist doch nur eine andere Form der Aussage. Die Beschreibung des (individuellen) Einzelsystems ist unvollständig!*
>
> *I am pleased to hear that you read my little article. Have you recognised also how illogical Pauli responded? He denies that this description is incomplete, but states in the same breath that the ψ function is a statistical description of the system, the description of a system-totality. This is just a different formulation of the statement: The description of an (individual) single-system is incomplete!*[28]

[28] Translation by the author.

This statement is the essence of *Born completeness*. And it shows that the EPR arguments and all its successor arguments by Einstein had one main purpose: To show the incompleteness of quantum mechanics for the specific purpose of justifying the need for a new field theory to replace quantum mechanics.

Despite the ambiguous notions of completeness, it is fair to conclude that it was the idea to present this new complete theory in the next published work following two months after the EPR paper; the Einstein-Rosen bridge paper.

We presented the meaning of the terms "completeness" and "reality" and pointed out which aspects of reality were specially relevant to Einstein in context with the 1935 papers. In the following chapter, we make use of these clarifications to discuss the connection of the 1935 papers in the context of the unified field theory program.

4.3.4 The Unified Field Theory Program

Einstein had in mind to resolve the incomplete[29] statistical character of quantum theory with a unified field theory. Even in 1934, one of the conclusive statements of his paper on the method of theoretical physics addresses the statistical character (Einstein 1934):

> *I still believe in the possibility of giving a model of reality, a theory, that is to say, which shall represent events themselves and not merely the probability of their occurrence.*

We have seen before that to Einstein the concept of completeness is strongly connected to the statistical character of quantum theory. In the original EPR argument in Einstein et al. (1935), he used the reality criterion, locality (Trennungsprinzip—separation principle), and the contradiction principle to conclude the incompleteness of quantum mechanics. In Einstein (1936), he no longer referred to those principles, instead, he argued for the ensemble interpretation, which is an indirect use of the *Born completeness*. According to the ensemble interpretation, quantum mechanics only refers to a statistical ensemble to events, not to a single event itself. To Einstein, this argument was sufficient to show the incomplete character of quantum mechanics and to head for an underlying theory. We can see that this is one of the main agendas of the EPR paper Einstein et al. (1935):

> *While we have thus shown that the wave function does not provide a complete description of the physical reality, we left open the question of whether or not such a description exists. We believe, however, that such a theory is possible.*

This new theory he presented in the ER-paper, at first independent of the EPR critique, was meant to solve the problem of the atomistic character of matter and fulfils Einstein's requirement of completeness (Einstein and Rosen 1935b):

---

[29] The completeness refers to *Born completeness*.

> *In favor of the theory*[30] *one can say that it explains the atomistic character of matter as well as the circumstance that there exist no negative neutral masses, that it introduces no new variables other than the $g_{\mu\nu}$ and $\phi_{\mu\nu}$, and that in principle it can claim to be complete (or closed). On the other hand one does not see a priori whether the theory contains the quantum phenomena.* [our underlining]

Here he means *theory completeness*, which is evident when he puts the term "closed" in brackets to underline that it refers to the closeness of a unified field theory that hopefully solves the particle problem.

It is remarkable that Einstein differs between the atomistic character of the theory and the question if the theory contains quantum phenomena. We have seen before what he means exactly by atomistic character", it refers to the problem of the localization of particles without accepting singularities of the field. The Einstein-Rosen bridge would have solved this in a way satisfactory to Einstein. It is not exactly clear what Einstein understands by the term "quantum phenomena" and how the theory could contain them. It is comprehensible that he means the quantum states and the formalism of quantum mechanics, which cannot be found in the Einstein-Rosen bridge representation for a particle, but it certainly was Einstein's hope that the quantum effects would emerge out of this new model.

In Einstein (1936), he presented the EPR argument in a modified form in line with the ER-bridge. Here, the connection of these two papers as part of a program is evident:

> *Es wird aber gezeigt, dass die Überzeugung von der Unfähigkeit der Feldtheorie, diese Probleme mit ihren Methoden zu lösen, auf Vorurteilen beruht.*
>
> *It will be shown that the conviction of the inability of field theory to solve these problems*[31] *rests on a prejudice.*

and

> *Angesichts dieser Sachlage erscheint es mir durchaus gerechtfertigt, die Frage ernsthaft zu erwägen, ob nicht doch die Grundlage der Feldphysik mit den Quanten-Tatsachen vereinbar ist.*
>
> *In view of this situation it seems appropriate to me to reconsider the question if a unification of the foundations of field theory and quantum facts is possible*[32]

This sums up the scientific program Einstein attempted to present. At the beginning, he refers to a comparison between quantum mechanics and statistical mechanics, and states that a statistical or effective theory is not an appropriate starting point to develop a complete[33] theory. In his view, quantum mechanics was this effective theory that delivers only statistical predictions, but gives no account for the single events.

---

[30] The Einstein-Rosen bridge model for elementary particles.

[31] The problems refer to the inability of field theory to give account for a molecular structure and to describe quantum phenomena with methods of field theory (Einstein 1936, p. 347).

[32] Translation by the author.

[33] In the sense of *Born completeness* and *theory completeness*.

In a more private environment, a letter from the 16th February, 1936 to his friend Michele Besso, Einstein expresses this point of view explicitly. Since to my knowledge this letter has not been cited in context with the EPR discussion, we want to cite it in full length[34] (Speziali et al. 1972, p. 308).

> *Ich halte die statistische Physik trotz all ihrer Erfolge doch für eine vorübergehende Phase und habe Hoffnung, zu einer wirklich befriedigenden Theorie der Materie zu gelangen. Ich sende Dir gleichzeitig eine kurze Arbeit, die den ersten Schritt darstellt. Das neutrale und das elektrische Teilchen erscheinen gewissermassen als Loch im Raume, derart, dass das metrische Feld in sich selbst zurückkehrt. Der Raum wird als zweischalig dargestellt. In der Schwarzschild' schen strengen zentralsymmetrischen Lösung erscheint das Teilchen im gewöhnlichen Raume als Singularität vom Typus* $1 - \frac{2m}{r}$. *Durch die Substitution* $1 - 2m = u^2$ *wird das Feld regulär im r-Raume. Wandert u von* $-\infty$ *bis* $+\infty$, *so wandert r von* $+\infty$ *zu* $r = 2m$ *und hierauf wieder zurück zu* $r = +\infty$. *So kommen beide "Blätter" im Riemann' schen Sinne zustande, die an der "Brücke"* $r = 2m$ *bezw.* $u = 0$ *stetig zusammenhängen. Aehnlich bei der Elektrizität. Die Aufgabe an der ich mit einem jungen Kollegen (russischer Jude) unablässig schwitze ist die Behandlung des Mehrkörperproblems auf dieser Basis. Wir haben aber die ernsthaften Schwierigkeiten des Problems bereits überwunden, sodass sich bald zeigen wird was daran ist. Jedenfalls ist es eine wundervolle mathematische Aufgabe.*

> *To me statistical physics despite its success is a transitory phase and I have hope that we arrive at a really satisfying theory of matter. I send you enclosed a short work, that represents a first step. The neutral and the electrical particle appear as a hole in space, of this kind that the metric field returns to itself. Space is represented as two sheets. In the strict spherical symmetric solution of Schwarzschild the particle appears in usual space as singularity of the kind* $1 - \frac{2m}{r}$. *By substitution* $1 - 2m = u^2$ *the field becomes regular in r-space. If u goes from* $-\infty$ *to* $+\infty$, *r is going from* $+\infty$ *to* $r = 2m$ *and back again to* $r = +\infty$. *This represents both "sheets", that are connected by the "bridge" at* $r = 2m$ *respectively* $u = 0$. *Likewise as it is in electricity. The challenge a young college (russian jew) and I labour away over is the many-body-problem on that basis. We have conquered the serious problems already, so it will show soon if there is something serious about it. Anyway it is a wonderful mathematical problem.*[35]

The connection between Einstein's critique on quantum mechanics, specifically its statistical character as well as the problems of the localizability of particles, and the connection to the Einstein-Rosen bridge approach, could not be closer than presented here.

Einstein's unified field theory program failed for reasons we will not discuss in this work. For a comprehensible presentation of this field theoretic program, refer to Van Dongen (2010). Einstein expressed his hopes how the program would succeed and how it might contain the quantum phenomena (Einstein 1936):

> *Erst die Untersuchung des Mehr-Brücken-Problems kann zeigen, ob diese theoretische Methode eine Erklärung für die empirisch erwiesene Massengleichheit der Teilchen in der Natur liefert, und ob sie den von der Quantenmechanik so wunderbar erfassten Tatsachen gerecht wird.*

[34] We omit a passage on the political situation in Europe.

[35] Translation by the author.

> *Only the examination of the many-bridge-problem can show if this theoretical model provides an explanation for the empirical proven equality of masses of particles in nature, and if it can reproduce the facts that are represented by quantum mechanics in such a delightful way.*[36]

These hopes of Einstein seem to be carried on in recent program claims that reconsider the idea of a deep connection between entanglement and Einstein-Rosen bridge solutions.

4.3.5 Recent Claims of EPR = ER

Recents claims bring back the EPR-ER idea in the context of Quantum Gravity (Maldacena and Susskind 2013):

> *General relativity contains solutions in which two distant black holes are connected through the interior via a wormhole, or Einstein-Rosen bridge. These solutions can be interpreted as maximally entangled states of two black holes that form a complex EPR pair. We suggest that similar bridges might be present for more general entangled states.*

and (Susskind 2016):

> *Now I feel that our current views of quantum mechanics are provisional; it's the best we can do without a much deeper understanding of its connection with gravity, but it's not final. The reason involves a very particular development, the so called ER = EPR principle. ER = EPR tells us that the immensely complicated network of entangled subsystems that comprises the universe is also an immensely complicated (and technically complex) network of Einstein-Rosen bridges.*

It is remarkable that the critique by Susskind sounds not that different from Einstein's reservations on quantum mechanics. The term "provisional" and the statement that quantum mechanics is not to be seen as final can be understood as synonymous to Einstein's claims for incompleteness of quantum mechanics, in the sense of *theory completeness*.

4.4 The EPR Paper Revisited

In the Appendix "The EPR paper—Refining the notions of completeness" we include a reprint of the 1935 EPR paper alongside colour-coded suggestions which concepts of completeness are used in the specific passages of the paper.

[36] Translation by the author.

We have used colour highlighting to mark the terms in the EPR-paper that refer to completeness and suggest their various used concepts of completeness. This illustrates the structure of the original EPR argument.

- *Theory completeness*—orange
- *Bijective completeness*—green
- *Schrödinger completeness*—blue.

Chapter 5
The Concept of Reality in the Interpretations of Quantum Mechanics

5.1 Reality in Classical Physics

Reality is an extremely broad concept, so we aim to provide an overview of its aspects and presumptions, as well as clarify its definitions for conceptual use. Since the conception of reality will change vastly in quantum mechanics, first we want to provide an overview on aspects of classical realism, which are relevant to our discussion on reality in quantum mechanics. As Braver (2007, p. 14) points out realism in not always defined in a straightforward way:

> *Of course, there are as many species of realism as there are kinds of objects, since realism can be type-specific or "local" rather than "global": one can be a realist about stars but not about the occurrence of the sequence "7777" in pi; a realist about the past, but not about the future; a realist about unobserved, but not unobservables; or the reverse for each pair.*

One could have the impression that "reality" is a far too broad conception to give a clear definition. Gallie (1955) provided a framework for terms that represent such a broad conception, the idea of Essentially Contested Concepts:

> *I shall try to show that there are disputes, centred on the concepts which I have just mentioned, which are perfectly genuine: which, although not resolvable by argument of any kind, are nevertheless sustained by perfectly respectable arguments and evidence. This is what I mean by saying that there are concepts which are essentially contested, concepts the proper use of which inevitably involves endless disputes about their proper uses on the part of their users.*

Essentially Contested Concepts usually refer to sociopolitical terms such as "democracy" and "freedom," but the term "reality" with its variety of interpretations gives the impression of being a good candidate for an Essentially Contested Concept. We do not examine in detail if "reality" conforms to the conditions of Essentially Contested Concepts, but want to adopt the spirit of the discussion on Essentially Contested Concepts for our examination of the term reality.

G. C. Krizek, *The Conceptions of Reality in the Interpretations of Quantum Mechanics*,
https://doi.org/10.1007/978-3-031-65786-3_5

5.1.1 The Structure of a Physical Theory and Its Metaphysics

We proposed a classification scheme to illustrate the formal, conceptual and ontological differences between various interpretations of quantum mechanics before. A visualization of the scheme shows the assignment of the different elements of a theory and the related metaphysics (Table 5.1).

On the mathematical level only mathematical symbols and their interrelations are introduced.

On level two of the classification scheme, names for the symbolically represented quantities are assigned. This is the introduction of the basis of a theoretical term. The full framework of theoretical terms is given by conceptual level three of the classification scheme. It contains principles and conceptions. Theoretical terms are fully defined on that level. An overview on the philosophical problems of theoretical terms is provided by Andreas (2016).

In the framework of the evolution of physical theories, the question on the status of theoretical terms arises again. Theoretical terms in several theories might be

Table 5.1 The classification scheme and its application

1		Mathematical laws between mathematical symbols	Physics
2	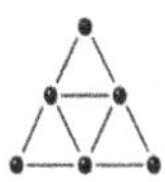	Interpretation physical quantities Relation physical quantities - Measurements Measurement laws	
3		Theoretical terms Concepts and principles	
4		Ontology - beables - elements of reality	Metaphysics

represented by the same semantical term, but the meaning might differ vastly as emphasized by Esfeld (2008, p. 30):

> *Der Wechsel von einer alten zu einer neuen Theorie kann derart schwerwiegend sein, dass die inferentiellen Beziehungen, die den Inhalt der charakteristischen Begriffe der beiden Theorien definieren, sich so radikal ändern, dass die charakteristischen Begriffe der alten Theorie nicht in den charakteristischen Begriffen der neuen Theorie formuliert werden können. Es gibt kein gemeinsames Maß für die charakteristischen Begriffe beider Theorien.*

> *The change from an old to a new theory can be radical in that sense, that the inferential relations, which define the content of the characteristic terms of booth theories, change radically in a way that the characteristic terms of the old theory cannot be formulated in words of the characteristic terms of the new theory. There is no common measure for the characteristic terms of both theories..*[1]

This concept is called Incommensurability and has been introduced by Kuhn (1962), Feyerabend (1993). In quantum mechanics the term "particle" does not have the same conceptual and ontological meaning as in classical particle mechanics.

Detailed accounts on the historical debates on the structure of scientific theories are provided by Suppe (1977), Sneed (1976), Ludwig (1978), Ludwig and Thurler (2007).

Following this brief overview on the structure of physical theories we provide a list of conceptions that contribute to the conception of classical realism.

5.1.2 Uniqueness

In classical physics we refer to old principles that build the very basis of scientific endeavour. This basis has been set by the Greek philosophers, especially Aristotle, who provided a logical system and a philosophy of nature.

Two of these basic principles influence our scientific thinking in a fundamental way: The contradiction principle and the Law of the Excluded Middle.

The contradiction principle is formulated in the following way (Ross 1928):

> *It is, that the same attribute cannot at the same time belong and not belong to the same subject and in the same respect.*

The contradiction principle can be seen as logical rule, as epistemological, or ontological rule. In that sense, it refers to different regimes of scientific thinking.

Depending on the method of interpreting Aristotles contradiction principle, it is related to a different level of the classification scheme. The scheme consists of four levels that represent the structure of a scientific theory: The mathematical formalism, the assignment to measurement values and theory-terms, the conceptual structure, and the ontology of the theory.

---

[1] Translation by the author.

The logical contradiction principle would refer to the mathematical formalism on level one, the epistemological[2] contradiction principle would refer to the level three of conceptions and principles of a theory, and the ontological contradiction principle would refer to the level four of the classification scheme.

In a formal way the contradiction principle is given by the statement on any statement p:

$$\neg(p \wedge \neg p) \tag{5.1}$$

The law of the excluded middle states that between a statement p and its negation $\neg p$ there is nothing in between. Therefore, in a formal way, the law of the excluded middle is given by

$$p \vee \neg p \tag{5.2}$$

These logical principles follow a priori from our common sense experience that objects are here or there, from the behaviour we observe in nature. This a priori knowledge influenced our thinking and the scientific method tremendously. So the state of existence of an ontological object is thought as unique, as unambiguous. This uniqueness is one essential basis of classical reality.

Quantum mechanics, with its concepts of superposition, entanglement, and non-classical statistics, challenges this view at the ontological level. Some interpretations, such as the Quantum logic approach challenged also the mathematical formalism and questioned the classical logical structure (Mittelstaedt 1978).

5.1.3 *Independence*

In classical physics, the conception of reality is independent of us, the observers. Reality was and will be there, independent if we describe it or not (Sankey 2008, p. 15ff). And the term "description" gives a good picture of the idea behind independence in this context. A description is the process of observation without altering the observed. It describes some reality that is there in a language, or another formal way. This independence will be realised within the definition of metaphysical realism in philosophy of science.

[2] Since the epistomological application is the one which is not straightforward, we want to give an example: In a political system like democracy, people are seen equal by principle. All citizens are allowed to elect. All non-citizens are not allowed to elect. Though we know that the individuals are not equal on an ontological level, in the conception of democracy an epistemological equality is adopted, which gives no implication to the ontological level for the individual human.

5.1.4 Completeness and Correspondence

Directly related to the notion of independence is the conception of correspondence and completeness. The idea of correspondence is that the scientific theories and their content corresponds to something out there. We observe phenomena and describe them with the formal language of mathematics and a semantic structure of terms and statements. It is one assumption of realism that this formal and semantic structure corresponds to a reality that is the causation of the observed phenomena (Putnam 1981; Hilary 1988).

The correspondence between reality and the formal and semantic description of the theory can involve different levels of the theory structure. We allude to the classification scheme and can distinguish pure mathematical correspondence and a conceptual correspondence. In a mathematical correspondence, the elements of the mathematical formalism of the theory correspond to elements on the ontological level. In the conceptual correspondence, concepts are corresponding to elements on the ontological level.

We provided a definition of different conceptions of completeness in Sect. 4.3.3. We summarize these conceptions of completeness and their relation to reality:

Theory Completeness

Definition 5.1 A theory fulfills theory completeness if its physical quantities and laws are connected by concepts that only relate only to each other and the theory renounces extra assumptions of concepts, which exceed the framework of the theory.

Bijective Completeness

Definition 5.2 Bijective completeness means that every element of the physical reality ϵ_i must have a counterpart t_i in the physical theory.

Born Completeness

Definition 5.3 Born completeness for a theory is fulfilled if the theory contains no probabilistic elements. A theory which contains statistical accounts and probabilistic statements is considered to be incomplete.

Schrödinger Completeness

Definition 5.4 A theory containing probabilistic statements about elements of the theory is Schrödinger complete if those statements refer to single events or entities and are fundamental in the sense that nothing more can be said about the system.

The theory is considered to be incomplete if there are potential new elements[3] in the theory that would abandon the probabilistic character of the theory.

[3] Einstein calls them "fremde Faktoren" (external factors). The translation of "fremd" is peculiar, since in the literal translation it would be "strange", "weird", "alien" or "different." But in context, it refers to factors that are not considered in the theory. They are not part of the theory, so we decided to use the non-literal translation "external." In our modern context, these factors could be understood as "hidden variables."

ψ -Completeness

An ontological model is ψ-complete if the ontic state space Λ is isomorphic to the projective Hilbert space $\mathcal{PH}$ (the space of rays of Hilbert space) and if every preparation procedure P_ψ associated in quantum theory with a given ray ψ is associated in the ontological model with a Dirac delta function centered at the ontic state Λ ? that is isomorphic to Ψ, $p(\lambda|P_\psi) = \delta(\lambda - \lambda_\psi)$.

The conception of completeness is deeply related to the conception of reality, since in a classical picture reality is mapped by correspondence rules to a physical theory. Since we expect this theory to describe the whole phenomena we consider it to be complete in the sense of *theory completeness* and by *bijective completeness* we expect it to map the whole of reality.

We conclude that *bijective completeness* is an inherent property of the conception of a classical realism. *Theory completeness* motivates the aim for a theory of everything which is fully complete in the sense of *theory completeness*, and by *bijective completeness* explores and defines all branches of reality.

5.2 Reality in Contemporary Philosophy of Science

A thorough discussion on reality in philosophy of science would go beyond the scope of this work, so we confine our presentation to conceptions that are already related to the debates on the interpretations of quantum mechanics. We also omit the historical development, since nearly every philosophical attitude advocated a position towards reality.

5.2.1 Metaphysical Realism

Metaphysical realism is the position that the world exists independently of us, our knowledge about the world and our perception of it.

We define a set of elements of reality PR.

$$PR = \{\epsilon_i\} = \{\epsilon_A, \epsilon_B, \epsilon_C\} \quad (5.3)$$

Definition 5.5 The world ensues Metaphysical realism if there exist ontological entities ϵ_i we name elements of reality.

The position of metaphysical realism is controversial and faces a lot of challenges, mainly based on the representation problem, which questions the connection between our beliefs and the independently existing entities in the world. The representation problem is the main attack line for criticism from anti-realists. There are several anti-realist positions that differ vastly, but their common ground is the rejection of

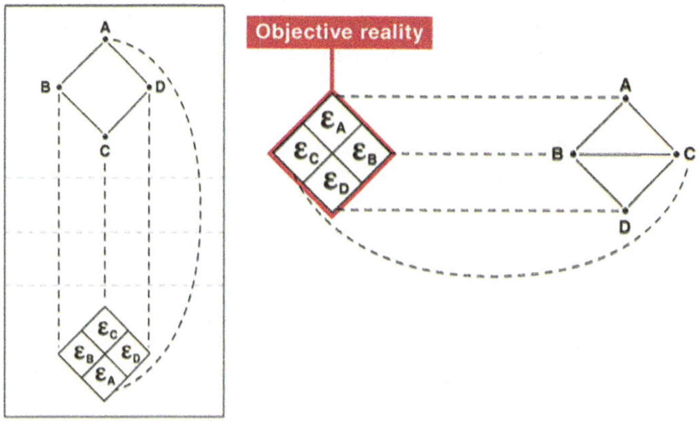

Fig. 5.1 Externalist metaphysical realism in the classification scheme

the basic assumption of metaphysical realism; the existence of a world independent of our perception and beliefs.

Metaphysical realism is an assumption prior to any statement on the connection to the structure of scientific theories. That is also the basic difference between metaphysical realism and scientific realism; the latter one already involves statements and positions towards the structure of our scientific enterprise. Metaphysical realism does not imply scientific realism (Khlentzos 2016):

> *That the world's constituents exist mind-independently does not entail that its constituents are as science portrays them. One could adopt an instrumentalist attitude toward the theoretical entities posited by science, continuing to believe that whatever entities the world actually does contain exist independently of our conceptions and perceptions of them.*

Putnam advances a position of internalist perspective of metaphysical realism and presents the externalist perspective in a brief way (Putnam 1981, p.48) (Fig. 5.1):

> *One of these perspectives is the perspective of metaphysical realism. On this perspective, the world consists of some fixed totality of mind-independent objects. There is exactly one true and complete description of 'the way the world is'. Truth involves some sort of correspondence relation between words or thought-signs and external things and sets of things. I shall call this perspective the externalist perspective, because its favorite point of view is a God's Eye point of view.*

It is noteable that Putnam in his presentation of the externalist view amends the basic assumption of metaphysical realism with the assumption of *Bijective completeness*.[4]

Additional to the set of elements of reality PR we define a set of elements of a theory T_{PR}.

$$T_{PR} = \{t_i\} = \{A, B, C\} \tag{5.4}$$

[4] See the definition for Bijective completeness.

Definition 5.6 The world ensues the externalist perspective of Metaphysical realism if there exist ontological entities ϵ_i we name elements of reality and there is one mapping ϕ that provides a bijection from every element ϵ_i to t_i.

Putnam clearly rejects this externalist view of metaphysical realism and advances a different position (Putnam 1981, p. 49–50):

> *I shall refer to it as the internalist perspective, because it is characteristic of this view to hold that what objects does the world consist off is a question that it only makes sense to ask within a theory or description. Many 'internalist' philosophers, though not all, hold further that there is more than one 'true' theory or description of the world. 'Truth', in an internalist view, is some sort of (idealized) rational acceptability - some sort of ideal coherence of our beliefs with each other and with our experiences as those experiences are themselves represented in our belief system - and not correspondence with mind-independent or discourse-independent 'states of affairs'.*

The internalist perspective advances the view that there is a metaphysical reality, but it is only possible only to conclude from the internal perspective of a theory to the nature of the elements of reality. In the internalist perspective, there is not one true and complete description of the world; there are probably more and even equally true descriptions that provide different mappings to the elements of reality. Therefore, we define the internalist perspective as the following:

Definition 5.7 The world ensues the internalist perspective of Metaphysical realism if there exists ontological entities ϵ_i we name elements of reality but there is no single mapping ϕ that provides a bijection from every element ϵ_i to t_i. There exist several sets T_{PR-k} and corresponding mappings ϕ_k from set T_{PR-k} to PR_k.

The assumption of metaphysical realism amended with statements on the structure of scientific theories represents the conception of scientific realism.

5.2.2 *Scientific Realism*

Scientific realism is based on metaphysical realism. There are several accounts on scientific realism, and we cannot provide a comprehensive overview of the field in this work. Thus, we adhere to the perspective outlined in Esfeld (2008):

> *Eine metaphysische Behauptung: Die Existenz und die Beschaffenheit der Welt sind unabhängig von den wissenschaftlichen Theorien.*
>
> ...
>
> *Eine semantische Behauptung: Die Beschaffenheit der Welt legt fest welche der wissenschaftlichen Theorien wahr sind (und welche nicht wahr sind).*
>
> ...
>
> *Eine epistemische Behauptung: Die Wissenschaften sind im Prinzip in der Lage, uns einen kognitiven Zugang zur Beschaffenheit der Welt zu gewähren.*

A metaphyiscal proposition: The existence and the nature of the world is independent of scientific theories.

...

A semantic proposition: The nature of the world determines which scientific theories are true (and which are not).

...

An epistemic proposition: Science empowers us to have cognitive access to the nature of the world.[5]

The first proposition is an addendum to the assumption of metaphysical realism, it is based on the existence of metaphysical realism and demands that scientific theories are independent of this nature of the world. The third proposition assumes that by our best theories we gain access to the metaphysical nature of the world, to the elements of reality.

The conception of scientific realism itself is also a matter of discussion to which we cannot refer in detail in this presentation. An overview on contrasting positions in this question is provided by Ellis (2005), Esfeld (2008), Psillos (2005), Sankey (2008).

A contrasting position to scientific realism is the position of Instrumentalism (Sankey 2008, p. 14):

Instrumentalism denies the literal interpretation of theoretical discourse, treating it instead as fictional discourse.

According to instrumentalism it is the aim of science to describe phenomena in a model. Every statement on metaphysical and specifically unobservable entities is superfluous in the framework of science. An instrumentalist's account on quantum mechanics is provided by Englert (2013):

Quantum theory has its interpretation and does not need a philosophical Überbau. Of course, philosophers should be encouraged to study the lessons of quantum theory - for example, the lesson that the future is not predetermined by the past - but their philosophical debates are really irrelevant for quantum theory as a physical theory, although they are likely to be of great interest to physicists. We could leave it at that, were there not the widespread habit of the debaters to endow the mathematical symbols of the formalism with more meaning than they have.

The discussion on the definition and relevance of scientific realism raises the question: If unobserved entities exist, what we can know about them? The most prominent discussion on the existence of unobserved entities in physical context is the atomic hypothesis: Do atoms exist? Or to be more precise: Is the conception of an atom within the physical theory necessary? And secondly: Does the conception of atoms imply that the existence of atoms as ontological objects is necessary or at least reasonable?

These two questions refer to different levels, the first only to the conception of a physical theory, the second to the ontological level according to the classification scheme. The question of Atomism should be seen also in the light of the old dichotomy between discrete and continuous.

[5] Translation by the author.

An historical introduction to Atomism with emphasis on Einstein's contributions is provided by Kox (2014), a review on the broader influences and implications of Atomism is provided by Roecklein (2015), and general introductions on Atomism are provided by Chalmers (2009, 2014).

The acceptance of atoms on the conceptual level was achieved relatively fast, the conception was too successful in the sense of unification and predictive power to be ignored. Therefore, the question arises: What are the principles that lead to the acceptance of new conceptions or ontologies?

This question addresses the distinction between physics and metaphysics which is a heavily disputed topic. van Inwagen and Sullivan (2017), Chap. 4 gives an overview on the problem and its recent status of the discussion:

> *One attractive strategy for answering these questions emphasizes the continuity of metaphysics with science. On this conception, metaphysics is primarily or exclusively concerned with developing generalizations from our best-confirmed scientific theories. For example, in the mid-twentieth century, Quine (1948) proposed that the "old/intermediate" metaphysical debate over the status of abstract objects should be settled in this way. He observed that if our best scientific theories are recast in the "canonical notation of (first-order) quantification" ..., then many of these theories, if not all of them, will have as a logical consequence the existential generalization on a predicate F such that F is satisfied only by abstract objects. It would seem, therefore, that our best scientific theories "carry ontological commitment" to objects whose existence is denied by nominalism.*

There is not one clarified and for all times fixed methodology to gain access to the metaphysical nature of reality; it is more a freedom of methods, which also was supported by Feyerabend (1993). One ontological conception which was both successful and comprehensible was the conception and ontology of objects.

5.2.3 Object Conception and Ontology

The conception of objects is not a trivial one, but for our discussion the following outline will be sufficient:

An object is an entity that has the ability to carry a set of properties and is distinguished from other objects by this set of properties. It is circumscribed from its environment.

These properties of objects are called intrinsic properties (Esfeld 2007, p. 7):

> *Intrinsic are all and only those properties that an object has irrespective of whether or not there are other contingent objects; in brief, having or lacking an intrinsic property is independent of accompaniment or loneliness.*

Properties that are not of that kind are called extrinsic or relational properties. They refer to relations to other objects.

This conception of objects is consistent with Einsteins view on physical objects (Einstein 1948):

> *It is further characteristic of these physical objects that they are thought of as arranged in a space time continuum. An essential aspect of this arrangement of things in physics is that they lay claim, at a certain time, to an existence independent of one another, provided these objects are situated in different parts of space.*

Since in classical physics objects like particles, rigid bodies, etc. play an important role in the ontology of these theories, a specific conception of reality is used, which is defined as Entity realism (Sankey 2008, p. 44):

> *Entity realism is therefore the doctrine that theoretical entities are real in the sense that they exist mind-independently. This may be expressed more simply as follows:*
> *(ER) The theoretical entities postulated by science are real.*

Not all theoretical entities will be regarded as real in the sense of Entity realism. It is a question of the structure of the theory, reason and practicability if a theoretical entity is considered as real. Since objects turn out to be conceptions in the best of our theories (within classical physics), it is reasonable to apply Entity realism to the conception of objects and regard them as real on the ontological level.

There are some arguments against an object ontology that follow directly out of the scientific progress in classical physics. At several states of classical physics and philosophy of nature before, objects on the ontological level have been proposed. During the progress of physics, these object ontologies have been abandoned or modified as long as physics arrived at atoms and elementary particles. All object ontologies that have been adopted on this way, together with respective sets of properties, turned out to have no ontological character. Instead, the properties assigned with those objects followed out further physical conceptions that involved more fundamental ontologies. Applying the Pessimistic Meta Induction to this development process, the object ontology turns out to be not a successful conception, though a helpful one. If object ontology fails on so many levels in classical physics, Pessimistic Meta Induction confined to this question questions why an object ontology should be applied to deeper levels. Would it not be surprising if object ontology failed on all empirical accessible levels and succeeds on the metaphysical level?

Object ontology is deeply related to materialism, since materialism emphasizes the fundamental character of matter as the essential entity of which everything is made. Materialism became challenged by theory of relativity and Atomism which deconstructed the conception of matter as a substance, a process that is continued by elementary particle physics even now.

The emphasis of objects in the conceptual and ontological level of our theories also led to several problems in context with quantum mechanics as we will see in Sect. 5.3. Since the object-conception and ontology is a dominant element in physical theories, we want to present a conception in philosophy of science that challenges the relevance of objects, the conception of structural realism.

5.2.4 Structural Realism

Structural realism is a conception in philosophy of science that goes back to Worrall (1989), who provided a conception for the structure of theories and the involved entities that differs vastly from the approaches in classical physics. The main difference is the type of theoretical entity that structural realism assigns reality to according to Entity realism. Structural realism emphasizes "structure" as the primary ontological entity and reduces the relevance of the entity "object."

(Worrall 1989, p.153):

> *Scientific Realism of any stripe consists of a metaphysical thesis conjoined with an epistemological one. The metaphysical thesis is often taken as the claim that there exists a reality independent of the human mind. But, as Elie Zahar points out, a more accurate rendition in our more enlightened, post-Cartesian-dualism times would be:*
> *There exists a structured reality of which the mind is a part; and, far from imposing their own order on things, our mental operations are simply governed by the fixed laws which describe the workings of Nature.*

and further:

> *Not only is this structured reality partially accessible to human discovery, it is reasonable to believe that the successful theories in mature science (the unified theories that explain the phenomena without ad hoc assumptions) have indeed latched on, in some no doubt partial and approximate way, to that structured reality, that they are, if you like, approximately true.*

With the structural realism approach, Worrall attempted to strengthen scientific realism, which had to face main criticism from the Underdetermination argument by Laudan (1981) and the Pessimistic Meta Induction by Quine (1951).

There are three positions of structural realism that can be distinguished, regarding their way of interpreting the Entity realism:

- Epistemic structural realism (ESR) is the position advocated by Worrall emphasizing the structures as primary over objects, but assigning both relevance on the ontological level.
- Radical Ontic structural realism (OSR) is the position introduced by Ladyman (1998b), emphasizing structure, but neglecting objects as superfluous, scrutinizing the metaphysical or epistemological aim of ESR.
- Moderate Ontic structural realism is the position advocated by Esfeld and Lam (2008) putting structures and objects on the same ontological basis. There are no relations without relata.

A more detailed overview on the differences of these positions is provided by Ladyman (2016). Despite all subtleties ESR and OSR are distinguished by these main positions (Ladyman 2016):

> *A crude statement of ESR is the claim that all we know is the structure of the relations between things and not the things themselves, and a corresponding crude statement of OSR is the claim that there are no "things" and that structure is all there is (this is called "radical structuralism" by van Fraassen 2006).*

Notwithstanding that object conceptions and ontologies played a crucial role in the conception of classical physics, in quantum mechanics the conception of objects becomes challenging and even dubious. We provide involved conceptions of reality in quantum mechanics and revisit the relevance of structural realism within quantum mechanics.

5.3 Concepts of Reality in Quantum Mechanics

The starting point for a new picture of reality is the new formal description of quantum mechanics that emerged out of the old quantum theory and its peculiarities. The formal structure of quantum mechanics was different in the sense of the involved framework. It was not defined in a $\mathbb{R}^3$ space or $\mathbb{R}^{1,3}$ Minkowski spacetime, it was formally defined on a Hilbert space, which has infinite dimensions. The formal structure of states as vectors in a Hilbert space allowed the possibility of superpositions.

With respect to that question, there are several conceptions of reality that are provided.

5.3.1 Naive Realism

Following the position of metaphysical realism presented in Sect. 5.2.1 there is a reality out there that causes the phenomena we observe. The view of naive realism is that our observations not only resemble the truth about the causes of the phenomena, even more our observations reveal the complete nature of these causes.

In these paradigms observation is a reading off of properties that are out there independently of our observations. Naive realism can involve realism about quantities as numerical representations of the properties of the entities in nature. Further, a naive realism about concepts is conceivable (Fig. 5.2).

For a presentation of Einsteins naive realism refer to Chap. 4.

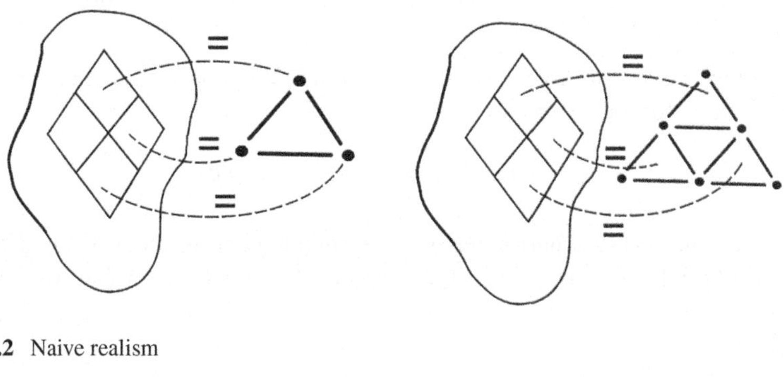

Fig. 5.2 Naive realism

A form of naive realism that is in discussion in contemporary physics is the Mathematical Universe Hypothesis (Tegmark 2004). It claims that all entities on the ontological level are directly related to the mathematical formalism, and even more are isomorphic to the mathematical formalism. In that sense, the Mathematical Universe Hypothesis trivially explains the unreasonable effectiveness of mathematics and proposed by Wigner (1960).

Naive realism is a version of the externalist perspective of Metaphysical realism:

Definition 5.8 Naive realism is the externalist perspective of metaphysical realism with either conceptual or mathematical elements of the theory being in *bijective completeness* to ontological entities, the elements of reality.

Naive realism involves no conception on measurements; therefore, it is not identical to a noninvasive measurability, which will be discussed later.

5.3.2 Einsteins Reality Criterion

In the EPR paper Einstein et al. (1935), Einstein presented a criterion for elements of a theory to represent elements of reality. The reality criterion is a methodological claim to identify elements of a theory that fulfil the requirement of being in correspondence with elements of reality.

Einstein was aware of the metaphysical demarcation problem with physics. In his letter to Schrödinger (von Meyenn 2010, letter 206, p. 537), Einstein admits:

> *Die eigentliche Schwierigkeit liegt darin, daß die Physik eine Art Metaphysik ist; Physik beschreibt "Wirklichkeit". Aber wir wissen nicht, was "Wirklichkeit" ist; wir kennen sie nur durch die physikalische Beschreibung!*[6]

> *The difficulty is due to that physics is a kind of metaphysics; physics describes "reality". But we do not know what "reality" is; we only know it through the physical description!*

The reality criterion is defined by Einstein et al. (1935) in the following way:

> *If, without in any way disturbing a system, we can predict with certainty (i.e., with probability equal to unity) the value of a physical quantity, then there exists an element of physical reality corresponding to this physical quantity.*

By definition of a probability measure $P(t_i)$ for the elements of the theory the reality criterion can be formally expressed as:

$$\forall t_i \in T_{PR} \wedge P(t_i) = 1 : \exists \epsilon_i \in PR \tag{5.5}$$

For Einstein the reality criterion has no deep ontological meaning or represents a general methodological claim; therefore, he got rid of it in later versions of the EPR

[6] Translation by the author.

argument. Fine (2009) claims that there are good reasons to believe that the reality criterion was an idea of Podolsky, since it is also incompatible with methodological aspects of Einsteins theory of relativity.

5.3.3 *Ruarks Reality Criterion*

In a response to the EPR-paper, Ruark proposed a modification of the reality criterion (Ruark 1935, p. 466):

> *This conclusion*[7] *is directly opposed to the view held by many theoreticians, that a physical property of a given system has reality only when it is actually measured, and that wave mechanics gives a faithful and complete description of all that we can learn from measurements. ... The reason*[8] *is that the authors adopt a criterion of reality different from the one just mentioned;*

And further:

> *The conclusion drawn by Einstein, Podolsky and Rosen is, that in accordance with their criterion of reality, the quantities corresponding to both P and Q possess reality; so that the wave description, failing to yield both of them, is incomplete.*

According to Ruark, none of both quantities correspond to an element of reality unless measured. This modified reality criterion resembles the position of Bohr, but without any commitment to classicality of the measurement apparatus.

5.3.4 *Local Realism in the Bell Inequalities*

Bell proposed an inequality in Bell (1964) that demonstrates the incompatibility of quantum mechanics with a set of assumptions in classical physics. These assumptions are locality and realism. We want to give a closer look to the definition of the latter one.

In its original intention Bell's theorem should demonstrate that only non-local hidden-variable theories like the De-Broglie-Bohm interpretation are not in contradiction with quantum mechanics, and all local hidden variable theories are ruled out, serving as an argument for the De-Broglie-Bohm interpretation (Bell 1964):

> *... a hidden variable interpretation of elementary quantum theory has been explicitly constructed. That particular interpretation has indeed a grossly nonlocal structure. This is characteristic, according to the result to be proved here, of any such theory which reproduces exactly the quantum mechanical predictions.*

---

[7] The conclusion that the quantum-mechanical description of the physical reality is complete.

[8] The reason for the divergence of conclusion.

It is irony that Bell's theorem later was used the other way around to argue that hidden-variable theories are untenable, and only by admitting that the De-Broglie-Bohm interpretation as a non-local hidden-variable theory, it was not affected by Bell's theorem.

Bell proposes the Bohm-version of the EPR argument with a pair of Spin-$\frac{1}{2}$ particles in a singlet state (Bell 1964):

> *Since we can predict in advance the result of measuring any chosen component of $\vec{\sigma}_2$' by previously measuring the same component of $\vec{\sigma}_1$, it follows that the result of any such measurement must actually be predetermined. Since the initial quantum mechanical wave function does not determine the result of an individual measurement, this predetermination implies the possibility of a more complete specification of the state. Let this more complete specification be effected by means of parameters λ.*

The parameter λ is the hidden variable that affects the value of the Spin vectors. In this classical model spin is assumed as a classical object property of the particles (Bell 1981, p. 1):

> *Do we not simply infer that the particles have properties of some kind, detected somehow by the magnets, chosen ala Bertlmann[9] by-the Source - differently for the two particles?*

Considering this, the conception of realism in Bell's theorem can be defined as:

Definition 5.9 Bell realism is the conception that a Bell-realistic theory describes objects with properties that are well-defined before the measurement process and after it.

Since the EPR argument was in the first place a way to criticise the statistical character of quantum mechanics by isolating the systems on hand of the principle of separability, quantum mechanics and, specifically the feature of entanglement, turned out to disobey the principle of separability. Soon the discussion on the EPR argument turned and separability was not longer an argument to isolate the measurement processes of the two constituents involved in a singlet state. Even more separability and locality itself were challenged. The only escape out of this situation from a classical realism perspective is to attribute correlation properties to the entangled constituents of the system (Bell 1981, p. 3):

> *To avoid such action at a distance they have to attribute, to the space-time regions in question, real properties in advance of observation, correlated properties, which predetermine the outcomes of these particular observations. Since these real properties, fixed in advance of observation, are not contained in quantum formalism, that formalism for EPR is incomplete.*

What Bell's theorem shows is that such an assignment of local pre-existing properties to objects does not reproduce the results as obtained from quantum mechanics. And the possibility to perform experiments to test if nature behaves according to these local pre-existing properties or according to quantum mechanics excels the value of Bell's theorem.

---

[9] Bell here refers to the Bertlmann's socks correlation as an example for a classical correlation.

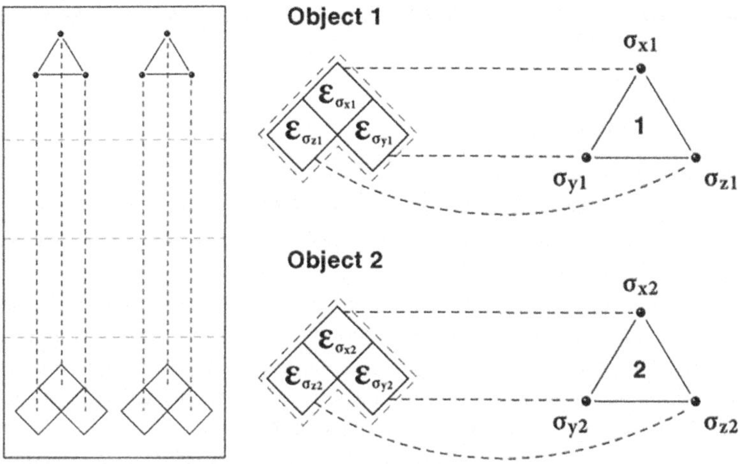

Fig. 5.3 Bell realism in the classification scheme

A huge number of experiments have been performed to test Bell's theorem and they convincingly demonstrated that nature behaves according to quantum mechanics (Clauser et al. 1969; Weihs et al. 1998). There are a number of loopholes in the conception of the experiments or the conception of Bell's theorem that could save local Bell-realistic theories. In the last decades, several of them were closed, but a full loophoole-free experiment has not been done.

Concluding, we can say that Bell-realism is defined by the object character of the quantum system carrying properties. In the classification scheme, this can be illustrated in the following way.

In a Bell-realistic theory, there are elements of the theory on level one and level two of the classification scheme. Corresponding to the elements of the theory, there are objects as entities on the ontological level that carry properties (Fig. 5.3).

5.3.5 *Reality in the Measurement Process*

In quantum mechanics, the measurement postulate is needed to explain how single eigenvalues arise out of the formalism, which describes a state generally in a superposition of several eigenfunctions and their according eigenvalues (Von Neumann 2018).

For classical physics, we want to propose a reality conception that describes the measurement process in classical terms.

Definition 5.10 Naive Measurement Reality is the conception that on the ontological level

(a) objects exist,
(b) that carry a set of properties,
(c) whose values are independent of the measurement process.

The difference between Bell realism and Naive Measurement realism is the ontological commitment to the existence of objects of the latter one. Bell realism does not involve such an assumption.

In the conception of Naive Measurement Reality, a measurement of the property of an object means that the value of the property of the object remains unaltered by the measurement process. The system that conducts the measurement changes one of its properties by the measurement process in a way that its value corresponds to or equals the value of the property of the measured system.

On the ontological level, the measured systems property is an element of the theory and corresponds to an element of reality; together they fulfil *bijective completeness*. By the measurement process, as explained above, one value of the measuring system will become corresponding or equal to the property of the measured system. By this measurement process, the property of the system that conducts the measurement alters one of its elements of reality. After the measurement, the measuring system will also fulfill *bijective completeness*.

The Naive Measurement realism proposes a noninvasive measurability, as the system that has been measured remains in its state unaltered, and a corresponding or equal value on the measurement apparatus evolves out of the measurement process.

If we apply this conception to the measurement process and the mathematical formalism in quantum mechanics, several problems arise. Gardner (1972) provides an analysis on the three positions of Bohr, Popper and Reichenbach and their relation to the Kochen-Specker theorem. We provide an overview of the reality conceptions in the picture of the classification scheme on these three positions.

5.3.6 Structural Realism in Quantum Mechanics

Already, Worrall (1989, p. 123) recognized the implications of the structural realism approach for quantum mechanics:

> *Is there any reason why a similar structural realist attitude cannot be adopted towards quantum mechanics? This view would be explicitly divorced from the "classical" metaphysical prejudices of Einstein: that dynamical variables must always have sharp values and that all physical events are fully determined by antecedent conditions. Instead the view would simply be that quantum mechanics does seem to have latched onto the real structure of the universe, that all sorts of phenomena exhibited by microsystems really do depend on the system's quantum state, which really does evolve and change in the way quantum mechanics describes.*

Or in a more compact way (Worrall 1989, p. 123):

> *The structural realist simply asserts, in other words, that, in view of the theory's enormous empirical success, the structure of the universe is (probably) something like quantum mechanical.*

Referring to the ontological level, structural realism rejects the idea of Bohr that statements and questions about the microscopical world are meaningless, but structural realism advances a realist view far from naive realism, related to scientific realism (Worrall 1989, p. 122):

> *The realist is forced to claim that quantum-mechanical states cannot be taken as primitive, but must somehow be understood or reduced to or defined in classical terms. But the structural realist at least is committed to no such claim - indeed he explicitly disowns it. He insists that it is a mistake to think that we can ever "understand" the nature of the basic furniture of the universe.*

What is emphasized by structural realism is the primary role of relations instead of a primary role of objects, but to which extent is a matter of dispute within the structural realist movement. Ladyman (1998a) first suggested that the position of Worrall (1989, p. 123) is ambiguous and introduced a distinction in epistemic and ontic structural realism, advancing the latter position. Worrals epistemic structural realism position is settled in the conception of Conservatism (Morganti 2004, p. 101):

> *... the proper interpretation of ESR is, I believe, exactly that according to which we are rationally entitled to believe in some level of preservation of structures (more or less faithfully representing reality) in the history of science but, considering the uncertainty existing at the level of our tentative ontological description of the universe, we should be 'agnostic' about what exists beyond those structures*

In Worrals view the objects as entities connected by structures are part of the structures, but not the primary ones. Ladymans positions is nowadays named as radical ontic structural realism and characterised by its full rejection of the concept of object entities, or individual entities (Ladyman 1998a, p. 422):

> *The demand for an individuals-based ontology may be criticised on the grounds that it is the demand that the structure of the mind-independent world be imaginable in terms of the categories of the world of experience. Traditional realism should be replaced by an account that allows for a global relation between models and the world, which can support the predictive success of theories, but which does not supervene on the successful reference of theoretical terms to individual entities, or the truth of sentences involving them.*

Related to this radical ontic structural realism, there is a position of moderate ontic structural realism advanced by Esfeld and Lam (2008) which is situated somewhere in between ontic and epistemic structural realism. The difference is again the ontological relevance of objects; to moderate structural realism, objects are an on the same footing as structures, in the sense that there are no relations without entities thate those relations refer to. In the context of quantum mechanics, moderate structural realism takes the following position (Esfeld 2007, p. 8):

> *Quantum physics can therefore be taken to suggest a metaphysics of relations, known also as structural realism: insofar as quantum physics is concerned, the fundamental physical properties consist in certain relations instead of being intrinsic properties.*

Moderate structural realism on one hand emphasizes the primary role of relations, but without abandoning the conception of objects in general (Esfeld 2004, p. 11):

> *The argument of this paper accepts that relations require things that stand in the relations (although these things do not have to be individuals, and they need not have intrinsic properties) and regards physical theories as referring to things.*

The merit of the radical OSR position following (Esfeld 2013, p. 29) is ...

> *... the commitment to there being a certain sort of structures in the domain of quantum physics instead of objects with an intrinsic identity.*

What Esfeld means exactly is that out of the formalism of quantum mechanics and its interpretation, and the framework of ontic structural realism, it is conclusive that there are no objects with intrinsic properties on the ontological level. What ontic structural realism is abandoning is the conception of objects as entities on the ontological level of the classification scheme, replacing it with the notion of structures.

What goes beyond this insight, is the claim that OSR cannot solve the problem without involving the interpretational positions of quantum mechanics (Esfeld 2013, p. 29):

> *OSR is not in the position to provide on its own an ontology for QM, since it does not reply to the question of what implements the structures that it poses.*

With this conclusion Esfeld (2013) presents an overview on three main candidates for interpretations of quantum mechanics that provide ontological implications: Many worlds interpretations, collapse interpretations, and hidden variable interpretations. Each of these interpretations provides a different candidate for the structure in the ontic structural realism program, and thereby completing OSR with an concrete structure ontology.

Esfeld (2013, p. 26) refers to Maudlin and presents three propositions for the measurment problem:

> *1.A The wave-function of a system is complete, i.e. the wave-function specifies(directly or indirectly) all of the physical properties of a system. 1.B The wave-function always evolves in accord with a linear dynamical equation (e.g. the Schrödinger equation).*
> *1.C Measurements of, e.g., the spin of an electron always (or at least usually) have determinate outcomes, i.e., at the end of the measurement the measuring device is either in a state which indicates spin up (and not down) or spin down (and not up).*

The three main interpretations according to Esfeld (2013) all give up one of these three propositions. Hidden variable interpretations give up 1A, collapse interpretations give up 1B and many worlds-interpretations give up proposition 1C.

What is omitted in Esfeld (2013) are several other approaches in the interpretations of quantum mechanics. In the following sections we want to amend the presentation of Esfeld (2013) by three more interpretational approaches that emphasize the conceptions of information and belief, but in completely different directions.

De Broglie-Bohm Interpretation

The De Broglie-Bohm interpretation is distinguished in essential aspects from Bohmian mechanics in the reading of Dürr and Teufel (2009). It is characterized by the view that the Schrödinger equation can be rewritten as a set of a complex- and a real-valued equation and resembles a Hamilton-Jacobi equation known from classical mechanics, supplemented by a term that has no correspondence in classical physics, the Quantum potential. Proponents of the De Broglie-Bohm interpretation emphasize neither the quasi-Newton character of the equations nor a return to classical physics. Even more, they emphasize the highly nonclassical character of the Quantum potential as an expression of a completely different physics. Therefore, it shows a deep misunderstanding of the De Broglie-Bohm interpretation if we follow the argumentation in Esfeld (2013, p. 28):

> *In enquiring whether OSR can be applied to Bohm's theory, let us focus on the contemporary conceptualization of this theory known as Bohmian mechanics, which abandons the quasi-Newtonian formulation of the theory in terms of a quantum potential*

For Bohmian mechanics, Esfeld (2013) identifies a candidate for the structure of a structural realism as:

> *it is a structure that takes all the particles as relata. In short, on Bohmian mechanics, there is a structure of correlated particle positions.*

In the De Broglie-Bohm interpretation this structure that takes all particle positions as relata in a very natural way, directly out of the reinterpretation of the Schrödinger equation, is the Quantum potential. For k particles it is defined as

$$Q = \sum_{k=1}^{N} -\frac{\hbar^2 \nabla_k^2 R}{2m_k R}. \tag{5.6}$$

The Quantum potential depends on a differential operator applied to the real-valued function R with the polar decomposition $\psi = Re^{i\frac{S}{\hbar}}$. The real-valued function R depends in general on the position of all k particles:

$$R = R(\vec{q}_1, ..., \vec{q}_k, t) \tag{5.7}$$

Thereby, the De Broglie-Bohm interpretation also instantiates the structure for a structural realist approach, but in the case of the hidden variable approach, in a more transparent way, with an explicit mathematical object, than in Bohmian mechanics.

Proponents of the De Broglie-Bohm interpretation emphasize the Quantum potential as an information potential, but with a different conception of information than in other informational approaches in quantum mechanics (Holland 1995, p. 90; Bohm and Hiley 1984, p. 260).[10]

It is information in the sense of in-formation,[11] not in the sense of knowledge of a system perceiving its environment (Bohm et al. 1987, 9.327):

> *In this explanation of the quantum properties of the electron, the notion of information plays a key role. Indeed it is helpful to extend this notation of information and introduce what could be called active information. The basic idea of active information is that a form having very little energy enters into and directs a much greater energy. The activity of the latter is in this way given a form similar to that of the smaller energy. It is therefore clear that the original energy-form will "inform" (i.e. put form into) the activity of the larger energy.*

In the picture of Maudlins propositions on the measurement problem, the De Broglie-Bohm interpretation gives up the proposition 1A, in the sense that it interprets the real-valued equation part of the Schrödinger equation as a Hamilton Jacobi function and, as a consequence of this, an additional equation of motion for the particle positions follows.

Proponents of the De Broglie-Bohm interpretation would identify active information and its mathematical carrier, the Quantum potential, as the resource that represents the structure that is requested by a structural realist approach.

Brukner-Zeilinger Interpretation

The Brukner-Zeilinger interpretation has been proposed as a foundational principle to reconstruct quantum mechanics out of informational principles (Zeilinger 1999).

For a presentation of the Brukner-Zeilinger interpretation refer to Brukner and Zeilinger (1999, 2003), Kofler and Zeilinger (2013), Zeilinger (1999). For critical remarks refer to Timpson (2003, 2013), Khrennikov (2016), Khrennikov et al. (2012).

Following the ideas of the Brukner-Zeilinger interpretation, information is seen as the fundamental entity on the ontological level.

Zeilinger (1999), p. 642:

> *The most fundamental viewpoint here is that the quantum is a consequence of what can be said about the world. Since what can be said has to be expressed in propositions and since the most elementary statement is a single proposition, quantization follows if the most elementary system represents just a single proposition.*

---

[10] It should be noted further that Bohm emphasized the preliminary character the Quantum potential has to him Bohm and Curd (1981, p. 102):

> *Nevertheless, we evidently cannot be satisfied with accepting such a potential in a definitive theory. Rather, we should regard it as at best a schematic representation of some plausible physical idea to which we hope to advance later, as we develop the theory further.*

[11] This reading of in-formation refers to Aristotle's formal cause.

By decomposition of complex systems into smaller and smaller parts one ends up at systems where such a single proposition represents the whole information content of the system. These type of systems are elementary systems according to Zeilinger (1999, p. 635).

Zeilinger (1999, p. 635) emphasizes that this foundational postulate can be seen in a rather instrumentalist view (Zeilinger 1999, p. 635):

> *I would like to underline that notions such as that a system "represents" the truth value of a proposition or that it "carries" one bit of information only implies a statement concerning what can be said about possible measurement results.*

To delimit this informational approach from subjective approaches like Qbism (Fuchs and Stacey 2016) or the Växjo interpretation (Khrennikov et al. 2012), Zeilinger emphasizes that information is not completely replacing the notion of reality, but he enhances the ontological status of information (Zeilinger 1999, p. 642):

> *It is evident that one of the immediate consequences is that in physics we cannot talk about reality independent of what can be said about reality. Likewise it does not make sense to reduce the task of physics to just making subjective statements, because any statements about the physical world must ultimately be subject to experiment. Therefore, while in a classical world-view, reality is a primary concept prior to and independent of observation with all its properties, in the emerging view of quantum mechanics the notions of reality and of information are on an equal footing.*

Contrary to the previously mentioned subjective approaches, the Brukner-Zeilinger interpretation makes use of the conception of a hypothetical observer to omit the assignment of the quantum state to one specific observer. This inter-subjective character of information and the description by quantum mechanics is emphasized by Brukner (2017, p. 6):

> *The quantum state is a representation of knowledge necessary for a hypothetical observer - respecting her experimental capabilities - to compute probabilities of outcomes of all possible future experiments.*

In context of David Deutsch's version of the Wigner's friend Gedankenexperiment, Brukner argues for the existence of facts only in a relative sense (Brukner 2017, p.1):

> *The view that holds the coexistence of the "facts of the world" common both for Wigner and his friend runs into the problem of the hidden variable program. The solution lies in understanding that "facts" can only exist relative to the observer.*

In that sense, reality and facts about the world are defined in a relational way only. This is supported by structural realism, which would emphasize this relational structure. Unclear remains the status of the quantum objects. The Brukner-Zeilinger interpretation remains neutral on the ontological structure of the objects on the quantum level, in a general sense, but it abandons the idea of classical pre-existing properties above the informational content of the elementary systems.

It is this informational storage capacity of an elementary system and its contained total information that is the candidate for a structure according to structural realist ideas. Entanglement, then, is the practical feature of quantum systems that share a structure of common or joint information.

One aspect of the Brukner-Zeilinger interpretation is the relation between the definition of information and the ways humans process information, as pointed out by Khrennikov (2016, p. 8):

> *I think that to explain peculiarity of the quantum representation of information one has to discuss the logical structure of information processing by humans. Besides classical Boolean logic, there exist various nonclassical logics, nonclassical rules for information processing. ... Meanwhile, the brain may use more complex logical systems. In particular, it may use quantum logic. Thus the quantum representation of information is a mathematical signature of the use of quantum logic in reasoning.*

Despite the inter-subjective approach in the Brukner-Zeilinger interpretation, Khrennikov emphasizes the subjective character of the whole process of information processing and the various definitions of logic. Contrary to the De Broglie-Bohm interpretation, where information is understood as a formal cause, information in the Brukner-Zeilinger interpretation is a still an entity that is related to observers.

QBism

A third approach which we want to relate to structural realism is QBism. QBism, earlier an abbreviation for Quantum Bayesianism, now renamed to Quantum Bettabilitarianism (Fuchs and Stacey 2016), is a continuation of the Copenhagen interpretation. It is a reconstruction of quantum mechanics that emphasizes specifically its subjective character, or merely the mixture of subjective and objective elements in one formalism. According to the proponents of QBism, this mixture is the source of the difficulties with the interpretation of the formalism of quantum mechanics (Fuchs 2017a, p. 8):

> *The starting point was, along with Jaynes, that the objective could never reside in the probabilities themselves; that was for the subjective. The objective must be somewhere else in the theory.*

A reconstruction of quantum mechanics based on SIC-POVMs, which are symmetric, informationally complete, positive operator-valued measures, is proposed to decompose the mixture of subjective and objective elements into a different formalism that emphasizes the probabilistic and subjective character of an agent interacting with nature. QBism speaks also about objective elements, which are identified as the Hilbert space dimension of the quantum systems (Fuchs and Stacey 2016, p. 47):

> *Quantum theory|that user's manual for decision-making agents immersed in a world of some yet to be fully identified character makes a statement about the world to the extent that it identifies a quality common to all the world's pieces. QBism says the quantum state is not one of those qualities. But of Hilbert spaces themselves, particularly their distinguishing characteristic one from the other, dimension,... QBism carries no such grudge. Dimension is something one posits for a body or a piece of the world, much like one posits a mass for it in the Newtonian theory. Dimension is something a body holds all by itself, regardless of what an agent thinks of it.*

QBism and the Brukner-Zeilinger interpretation agree in the emphasis of the significance of the Hilbert space dimension, but disagree on the interpretation of the very nature of this characteristic. Brukner and Zeilinger assign the conception of information to the elementary system, whereas QBism speaks more generally of a capacity (Fuchs and Stacey 2016, p. 48):

> *The claim here is that quantum mechanics, when it came into existence, implicitly recognized a previously unnoticed capacity inherent in all matter call it quantum dimension.*

By the emphasis of the Hilbert space dimension, QBism follows an old methodology of philosophy and physics equally; the methodology of identifying invariants as foundational and ontological entities. This invariant is the Hilbert space dimension or quantum dimension of a quantum system.

Following the classification of Cabello (2015) on interpretations of quantum mechanics, QBism is classified as a Type II interpretation on belief. If we want to relate QBism to structural realism, neither ontic structtural realism, nor epististemic structural realism seems suitable to represent the position of QBism, since both structural realist positions do not consider the subjective perspective of an agent in the world as emphasized by QBism.

Therefore we propose QBism as a different form of structural realism, a doxastic structural realism. Doxastic is derived from the greek word doxa, which stands for "belief", and reflects the distinction from epistemic and ontological commitments. The structures of this doxastic structural realism, that is represented by QBism, are the mutually beliefs of agents performing quantum mechanics. The structures identified by QBism are not objective, since they represent an internalist view of the beliefs on the behaviour of nature of one specific agent in interaction with nature.

The Qbist position also alters the definition of physics its purpose (Fuchs and Stacey 2016, p. 47):

> *What is being realized through QBism's peculiar way of looking at things is that physics actually can be done without any accompanying vicious abstractionism. You do physics as you have always done it, but you throw away the idea everything is made of [Essence X]" before even starting. Physics in the right mindset is not about identifying the bricks with which nature is made, but about identifying what is common to the largest range of phenomena it can get its hands on.*

So besides the subjective beliefs of agents, QBism identifies the quantum dimension as some objective structure that can be identified as common to a large range of phenomena. In our view this property reflects some aspect of the nature of the universe.

The identification of the quantum dimension as a capacity or resource entails a consequences. One has to deal with the infinite dimensional Hilbert spaces. Fuchs and Stacey (2016, p. 52) speculate that the quantum dimension should be finite in general:

> *To make quantum dimension meaningful in ontic terms, as a quality common to all physical objects, is to say it should be finite - going up, going down from this object to the next, but always finite. Every region of space where electromagnetism can propagate, finite. Every region of space where there is a gravitational "field," finite.*

Space holds an exceptional position among all degrees of freedom in quantum mechanics, and the solution to this conundrum likely lies beyond the framework of quantum mechanics. This exceptional status of space is evident through various interpretations of quantum mechanics. Using the example of Bohmian mechanics in the reading of moderate structural realism, the position of a particle remains the only intrinsic property left.[12] All other properties turn out to be relational and support the idea of an abandoning of the object-character of quantum systems.

5.4 Ockham's Razor and the Interpretations of Quantum Mechanics

The struggle over the interpretation of quantum mechanics has a long tradition, beginning in the 1920s, and often involves debating principles. One principle we discuss here, which emerged relatively late in interpretational discussions, is Ockham's razor. The emergence and use of this principle suggests that the discussion had stagnated due to a lack of new experimental evidence, entrenched positions, and the influence of a new generation of physicists.

In the following, we present a collection of citations from papers and publications where Ockham's razor was applied. We first present a short overview of the interpretations involved, present the citations by order of their positions, summarize the views and draw our conclusions.

5.4.1 Overview of Interpretations Used in the Discussion

Copenhagen Interpretation

The Copenhagen interpretation represents the view that the wavefunction, or the state vector, of a system gives a complete description of the physical situation of a quantum mechanical system. The evolution of the wavefunction or state vector is given by the Schrödinger equation. The measurement process of a physical quantity is not described by the Schrödinger evolution, it is defined by the projection postulate which assigns measurement outcomes to eigenvalues of specific operators. This process is often called the collapse of the wavefunction, because during a measurement process the unitary evolution of the wavefunction is replaced by this non-unitary collapse. According to Copenhagen interpretation, this collapse occurs objectively random and the physical quantities have no definite values before the measurement process. Furthermore, according to the Copenhagen interpretation, it has no meaning at all to speak about the values of a physical systems quantities prior to the measure-

[12] It is remarkable that Bohmian mechanics follows the position of moderate structural realism insofar as there are objects on the ontological level but with a lack of intrinsic properties.

ment. The Copenhagen interpretation is therefore a indeterministic,[13] non-realistic[14] interpretation of quantum mechanics.

De Broglie-Bohm Interpretation

The De Broglie-Bohm interpretation is a realistic, deterministic, nonlocal[15] hidden-variable interpretation and is based on the fact that the Schrödinger equation can be interpreted as set of one real- and one complex valued equation. The complex-valued equation is interpreted as the conservation of probability, and the real-valued equation is interpreted as a Hamilton Jacobi equation, with an extra term. This term is interpreted as a Quantum potential. From the Hamilton Jacobi equation an equation of motion for the particles can be derived, which is called guidance equation. In the De-Broglie Bohm interpretation, nonlocality is an important and inherent feature of quantum mechanics.

Bohmian Mechanics

Advocates of Bohmian mechanics claim that it is an independent theory based on the ideas of the De Broglie-Bohm interpretation, but it postulates the guidance equation for particles. In that view, it is not seen as an interpretation of quantum mechanics, and the guidance equation is not seen as a consequence of the Schrödinger equation. Bohmian mechanics claims to be identical in all experimental predictions with standard quantum mechanics and is a realistic, deterministic, nonlocal hidden-variable theory. Its fundamental entities are the particles and their positions.

Many Worlds Interpretation

The Many Worlds interpretation of quantum mechanics is a realistic deterministic interpretation that claims that the wave function is a complete description of the quantum mechanical processes. Therefor no collapse of the wavefunction hypothesis is necessary to explain the measurement process. During the measurement process a specific branch of the wavefunction emerges through the measurement interaction.

---

[13] Indeterminism is the position that an event has no causal reason; it is the opposite of determinism, where every event has causal reason. Both positions are connected to the concepts of probability and free will.

[14] In this context, realism means that properties and quantities of physical systems have a meaning and definite value prior to their measurement. Nonrealistic interpretations would deny this assumption and claim that it makes no sense to speak about quantities of physical systems prior to measurement.

[15] Locality is the concept that events are independent of all events that are space-like separated. Nonlocality abolishes this restriction and, therefore, events can have an instantaneous influence on space-like separated events.

QBism

QBism or Quantum Bettabilitarianism (Fuchs and Stacey 2016) is a continuation of the Copenhagen interpretation. It is the attempt to reconstruct quantum mechanics on the basis of probabilities as representation of the subjective degrees of belief of agents interacting with the world. The quantum state is not seen as a foundational entity, it is merely a different way to represent the probabilities. The Hilbert space is constructed out of SIC-POVMs which are symmetric, informationally complete, positive operator-valued measures. They generate the probabilities for the individual agents based on their subjective beliefs according to the fundamental postulate that reproduces the Born rule for the probabilities.

QBism is a local theory, takes a realistic position (for a detailed account on participatory realism, refer to Fuchs 2017b) and provides an epistemic interpretation of the wavefunction.

Mathematical Universe Hypothesis

The MUH is a hypothesis and not an interpretation of quantum mechanics, but it is connected to the Many Worlds interpretation, since the Many Worlds interpretation is contained in the MUH. The MUH claims that mathematical objects form the only external reality. It naturally explains Wigner' Unreasonable Effectiveness of Mathematics (Wigner 1960).

5.4.2 In Favor of de Broglie-Bohm Interpretation and Bohmian Mechanics—Against Copenhagen Interpretation

Detlef Dürr (Esfeld 2012, p. 116) argues in favour of Bohmian mechanics and against the application of Ockham's razor contra Bohmian mechanics. By standard quantum mechanics, he means the Copenhagen Interpretation.

> *One could say, that all additional things in Bohmian mechanics, the particles and it's trajectories, should fall victim to Ockham's razor. We believe this is a wrong conclusion. We believe that ontology is nothing secondary, contrary it should be at the beginning of a theory, and that Ockham's razor leaves behind a mutilated incomplete description of things.*

or

> *On one hand you have to accept that the description of measurements need rules that are completely different from all other laws even and cannot be derived from them. Already here fails the argument of Ockham's razor, standard quantum mechanics is not more economic than Bohmian mechanics.*

Lucien Hardy argues in favour of hidden variables (Hardy 2013):

It is very often argued that Ockham's razor should be applied here to rule against introducing such hidden variable, but this rather depends on how one interprets the objective of applying Ockham's razor and the need to provide explanation. There are perhaps three reasons why one might want to introduce hidden variables into quantum mechanics: (1) to restore determinism (2) To provide a clear ontology (3) To solve the measurement problem.

Oliver Passon indicates in Passon (2004) that Ockham's razor is used in the wrong direction and would actual favour Bohmian mechanics:

...the deBroglie-Bohm theory supplements ordinary quantum mechanics by an equation-of-motion for the quantum-particles, but eliminates the postulates which are related to the measurement process(not to mention how uncompelling those postulates are).

and further

Hence, it is questionable whether the precondition for applying Ockham's razor is met.

It is unclear what precondition Passon here refers to. The message still is clear; to him, Ockham's razor cannot decide in favour of Copenhagen interpretation.

5.4.3 In Favor of Copenhagen Interpretation—Against de Broglie-Bohm Interpretation and Bohmian Mechanics

One more example of how Ockham's philosophy is used to argue in favour or against a certain interpretation comes from Anton Zeilinger (2003, p. 151,152,154) when he speaks about Ockham's razor:

It is the assumption, that one should not invent things, quantities or entities without necessity.

and

William of Ockham was a medieval philosopher who wanted to cut away all unnecessary things in philosophy.

and

An important argument against the quantum potential comes again with Ockham's razor. If we can, what is indeed the case, explain just as much as without the quantum potential, then it is redundant.

Zeilinger (1996) furthermore argues in favour of Copenhagen interpretation:

I have purposely not dealt with questions like: Is there a border between micro- and macro physics? Is a new form of logic necessary for quantum processes? Has one's awareness an active, dynamic influence on the wave function? Such or similar positions were proposed by several physicists, but in my opinion they would all fall victim to Occam's razor: Entia non sunt multiplicanda praeter necessitatem. It is the beauty of the Copenhagen interpretation that it operates with a minimal set of entities and concepts.

5.4.4 In Favor of Copenhagen Interpretation—Against Many Worlds Interpretation

Anton Zeilinger applies Ockham's razor also against the Many World Interpretation (Zeilinger 2003, p. 151,152,154):

> *The Many-worlds interpretation violates a basic assumption, which has been very successful in the past. Why should one invent so many worlds, if there are other interpretations, which do the same without them?*

Nicolas Gisin (2013, p. 5) argues against Many Worlds, but unfortunately the reference in the paper refers not to the resource of the original critique:

> *Years ago, I once argued that the many-worlds does not seem compatible with Occam's razor principle. As answer I got the following: "Occam's razor should not be applied to the physical world, but be applied to the Schroedinger equation; don't add any term to this beautiful equation". The linearity of the Schroedinger equation was assumed more real than our physical universe.*

5.4.5 In Favor of Many Worlds Interpretation—Against de Broglie-Bohm Interpretation and Bohmian Mechanics

Hilary Greaves argues with Ockham's razor in favour of Many Worlds Interpretation against the hidden variable approaches, which adresses the De Broglie-Bohm Interpretation and Bohmian mechanics (Greaves 2007):

> *The motivations for the many-worlds interpretation include (i) the fact that (if defensible) it solves the measurement problem, (ii) compatibility with the spirit of special relativity, and (iii) Ockham's razor. ... There is no 'ontological extravagance' in any offensive sense, since nothing is being added at the fundamental level. (Indeed, the fundamental ontology of this sort of many-worlds interpretation is actually more parsimonious than that of the rival 'hidden variables' approaches.*

Vaidman (2016) argues in favour of the Many Worlds interpretation against Bohmian mechanics:

> *The MWI is also more economic than Bohmian mechanics, which has in addition the ontology of the particle trajectories and the laws which give their evolution.*

5.4.6 In Favor of de Broglie-Bohm Interpretation and Bohmian Mechanics—Against Many Worlds Interpretation

Valentini (2010) in a reply to claims that the Many Worlds interpretation includes the idea of De Broglie-Bohm interpretation:

However, since the de Broglie velocity field is single-valued, trajectories $q(t)$ cannot cross. There can be no splitting or fusion of worlds. The above de Broglie-Bohm multiverse then has the same kind of 'trivial' structure that would be obtained if one reified all the possible trajectories for a classical test particle in an external field: the parallel worlds evolve independently, side by side. Given such a theory, on the grounds of Ockham's razor alone, there would be a conclusive case for taking only one of the worlds as real.

5.4.7 Other Positions

The following citations cannot be assigned clearly to one of the previous groups, but are relevant to our discussion. Specifically, the Mathematical Universe Hypothesis is related to the Many Worlds interpretation. In our analysis we assign them to the same group and merge their points of view.

Arthur Fine argues with a counterfactual gedankenexperiment on the development of quantum mechanics (Hardy 2013, Arthur Fine—On the interpretation of Bohmian mechanics):

Given the entanglement of physics with positivistic philosophy in the first half of this century, a process initiated by Einstein and relativity, we might embellish Cushing's counterfactual history a bit. We might conjecture that if Bohmian mechanics had indeed been the first theory of the quantum domain, subsequent epistemological discussions might well have inspired at least some participants to wield Occam's razor to pare away everything that is not simply quantum mechanics!

Jannes (2009) with a critical remark on the Max Tegmark's Mathematical Universe Hypothesis (MUH):

A classical argument against Platonism is that it unnecessarily complicates our view of reality by requiring a commitment to the existence of an immense realm of mathematical and other abstract entities, thereby violating Occam's razor. This does not imply a denial of the effectiveness of mathematics at describing (part of) reality.

Tegmark (2004) with a statement concerning his Mathematical Universe Hypothesis and the application of Ockham's razor to the Multiverse concepts. He defends the Multiverse concept against Copenhagen interpretation and along the way attacks the simplicity of the Copenhagen interpretation:

In short, the Level III multiverse, if it exists, adds nothing new beyond Level I and Level II just more indistinguishable copies of the same universes, the same old storylines playing out again and again in other quantum branches. Postulating a yet unseen nonunitary effect to get rid of the Level III multiverse, with Ockham's Razor in mind, therefore would not make Ockham any happier.

and

The principal arguments against parallel universes are that they are wasteful and weird, so let us consider these two objections in turn. The first argument is that multiverse theories are vulnerable to Ockham's razor, since they postulate the existence of other worlds that we can never observe. Why should nature be so ontologically wasteful and indulge in such

> *opulence as to contain an infinity of different worlds? Intriguingly, this argument can be turned around to argue for a multiverse. When we feel that nature is wasteful, what precisely are we disturbed about her wasting? Certainly not space, since the standard flat universe model with its infinite volume draws no such objections. Certainly not mass or atoms either, for the same reason once you have wasted an infinite amount of something, who cares if you waste some more? Rather, it is probably the apparent reduction in simplicity that appears disturbing, the quantity of information necessary to specify all these unseen worlds. However, as is discussed in more detail in Tegmark (1996), an entire ensemble is often much simpler than one of its members.*

Fuchs (2010) remarks concerning a specific aspect of QBism:

> *The aim of physics is to find characteristics that apply to as much of the world in its varied fullness as possible. However, those common characteristics are hardly what the world is made of - the world instead is made of this and this and this.*

This statement does not address Ockham's razor, but it points out the position of Ockham in the debate on universals; it is a nominalists point of view. It is worthwhile to analyse if the characteristic of a position in the debate on universals can be found in other interpretations as well. The scope would be to answer whether it would take a realist position, which means in the context of the problem of universals, a neutral position, and if this characteristic can be used as a differentiation between interpretations.

5.4.8 *Analysis of Applications of Ockham's Razor*

The presented meta-analysis cannot claim to be comprehensive, as it did not cover all journals or a significant subset of relevant books in this field. While it cannot be used to draw quantitative conclusions, it does provide a representative overview of the current status in the literature, reflects a notable situation, and allows us to draw meaningful conclusions.

In the above meta-analysis, Ockham's razor is used in nearly all constellations by proponents of all positions. Positivist against realist point of view, realist against positivist point of view, deterministic interpretation against indeterministic interpretation, orthogonal interpretations arguing against each other. Figure 5.4 shows the positions and their respective contradictions. It turns out that they are all in contradiction to each other.

How is it possible that a principle that is meaningful to several advocates of different viewpoints in the interpretational debate allows such contradictory conclusions?

One of the problems of the application of Ockham's razor in general is that the principle itself argues in favour of parsimony over plentifulness, in favour of simplicity. And there it becomes difficult in our application to the interpretations of quantum mechanics. As we have seen, the definition of simplicity is not a standard one; even more in the above-cited viewpoints, it is rarely argued what is meant by this simplicity, as if it was obvious. A resolution for this could be to define a complexity measure of theories and then let the more simple theory according to this measure

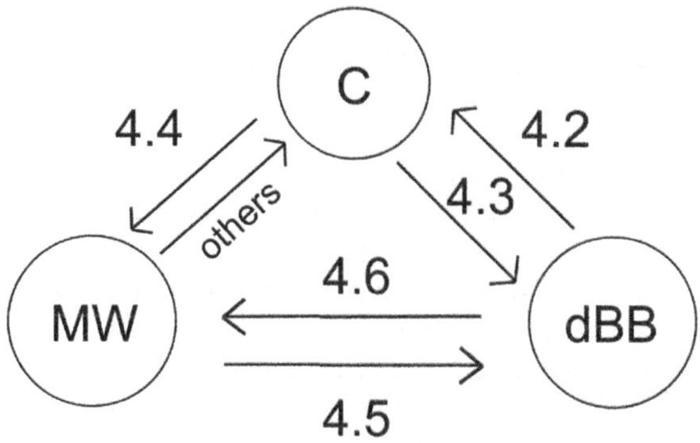

Fig. 5.4 Contradictions in application of Ockham's razor

(or in our case of interpretations of quantum mechanics, the more parsimonious interpretation) pass. It seems that the problem then only shifts to the definition of this measure of complexity, which then will become the heavily disputed topic. One could think that this leads into an endless self-referring circle with simplicity measures that refer to simplicity. On the other hand, a measure for the complexity of an interpretation is an interesting open question that should not be unconsidered.

5.4.9 Ockham's Razor and Economy of Thought

A different approach to this topic would be if we revisit the original idea of Ockham's razor as something of common sense, as an idea of economy of thought. This idea can be found in social psychology as well.

Wegener and Petty (1998) gives a review of the social psychology model of the naive scientist (Heider et al. 1958). The idea is that humans perceive their environment like naive scientists and make attributions to phenomena.

The naive scientist model got under attack over the years and has been replaced by the cognitive miser model, which argues that humans make indeed make use of several heuristics on their judging on certain facts, but the focus lies on the efficiency of the decision process (Fiske and Taylor 2013):

> *Hence the third general view of the thinker is the cognitive miser model. The idea is that people are limited in their capacity to process information, so they take shortcuts whenever they can. People adopt strategies that simplify complex problems; the strategies may not be correct or produce correct answers, but they emphasize efficiency. The capacity-limited thinker searches for rapid, adequate solutions rather than for slow, accurate solutions.*

Tversky and Kahneman (1973) remark on the methods of application of the different heuristics:

> *Little is known, however about the frequency of classes or the likelihood of events. We propose that when faced with the difficult task if judging probability or frequency, people employ a limited number of heuristics which reduce these judgements to simpler ones.*

and

> *We suggest that in evaluating the probability of complex events only the simplest and most available scenarios are likely to be considered.*

We see a deep connection between the cognitive miser model in social psychology and Ockham's razor in the way Ockham understood it, as a concept of common sense, as an economy of thought.

We conclude that the remarkable situation of contradictory statements concerning the simplicity of interpretations is due to the in-applicability of Ockham's razor in the interpretations of quantum mechanics. This in-applicability reflects in the contradictions between the different viewpoints in the interpretations of quantum mechanics, which shows that there is no preferred direction in how Ockham's razor is used, and it seems that it is up to subjectivity as to how it is applied. It therefore seems arbitrary how Ockham's razor is applied here at all, which allows doubts for the value Ockham's razor can have in the discussion on the interpretations of quantum mechanics.

As long as there is no quantitative system that allows to account for simplicity in that regime, we are forced to rely on subjective judgements if Ockham's razor is used reasonably in that specific case. We think that this is not an appropriate scientific method, but merely a matter of taste. We should be very careful and alert when applying Ockham's razor, especially if we adopt it as a first order principle, since the analysis of its historical origin showed that Ockham's razor is just a concept of economy of thought and not a concept of simplicity of nature, which is also supported by social psychology.

5.4.10 A New Diversity

We have come a long way from the medieval philosophy and its peculiarities to the disputes on identity, which are still an open question in the diverse field of the interpretations of quantum mechanics. We questioned the applicability of Ockham's razor to the interpretations of quantum mechanics and concluded that no clear scheme can be seen as to how Ockham's razor should contribute in a positive way to the ongoing discussions. This is certainly owed to the notion of simplicity, which is merely a subjective quality and seems to be a relative concept.

Alluding to the Great Cosmological Controversy again, we claim that there is an analogy[16] to the current situation in the interpretations of quantum mechanics. In the case of the cosmological models, we encounter two orthogonal world-views, but overlapping in the empirical predictions of the models[17] that rival for being accepted as the "true" theory. As argued before in the discussion on the Ontological

[16] Though there are clearly different situations concerning the mathematical framework.

[17] With respect to the experimental limitations of that time.

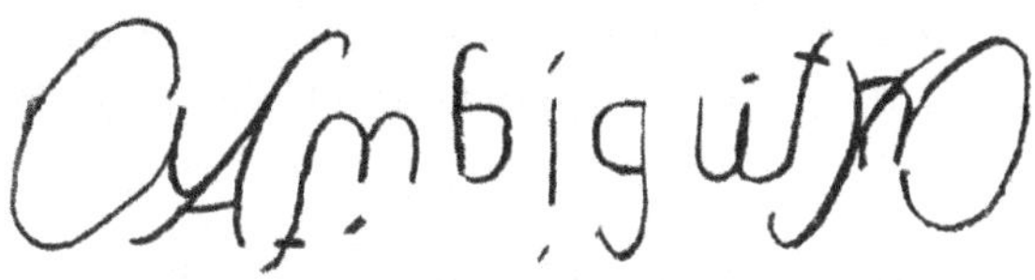

Fig. 5.5 Quambiguity

Coherence Argument, each position claimed simplicity as an argument for its world-view, underpinning it with significant arguments.

It turned out that the heliocentric model delivered a world view that had striking advantages, but when relativity arose at the begin of the 20th Century the Great Cosmological Controversy was seen under a new perspective. The remarkable situation that it was not a trivial task to decide between these two models turned out to have a reason, in the fact that both models are applicable due to the relativity principle, and the question for truth if the sun or the earth rested turned out to be meaningless. In quantum mechanics, we are in the exceptional situation that we face several interpretations that are in full agreement with the experimental predictions but are completely orthogonal in their ontological statements and implications, even with respect to diverse aspects of the world views.[18] We conclude out of this situation in the field of the interpretations of quantum mechanics that we have an analogy to the Great Cosmological Controversy.

Is this situation not an indication that should encourage us to think that a dogmatic dispute with one winner might not be the solution of the problem? One can believe that the current situation in quantum mechanics can be understood that way and that this situation can be interpreted as a perspective on the new physics that may lie ahead, indicating that our well-established concepts of quantum-objects, spacetime, determinism, probability and information will be challenged. Even more, it seems that our philosophy of science will be challenged. Quantum logic showed that even the propositions of logic can be involved in the ambiguous story of quantum mechanics (Mittelstaedt 1978). To underpin the importance of this ambiguousness, we call this quality of quantum mechanics "Quambiguity" (Fig. 5.5).

It has been shown by Quine (1951) and others that there is an underdetermination of science in general, but specifically in quantum mechanics it seems that this underdetermination is raised to a new level. Quantum mechanics allows a variety of interpretations that advance completely orthogonal ontologies. Is this "Qambiguity" a feature of the formalism, and therefore a question of the representation of the empirical facts? Or, on the other hand, does "Quambiguity" give an insight into the ontological structure of the world beyond quantum mechanics, and what is the conclusion if it is possible to involve such a variety of ontological conceptions with one formalism and one set of empirical facts?

We cannot give an answer on these serious questions, but we can give some conjectures. It seems possible again to take a realist and anti-realist position on the

[18] Such aspects of interpretations which show orthogonal behavior are Determinism-Indeterminism, Locality-Nonlocality, Nominalism–Platonism, Objectivity–Subjectivity, Uniqueness–Ambiguity.

question on the nature and reason for Quambiguity. A anti-realist position could be to assert that the ambiguity in the ontological commitment is an indicator for the redundancy of any ontological commitment. This could be the position of a radical Empiricism, or some related view. A realist position could be to question the nature of ontological entities that we rely on. Usually, realism is related to an object-realism, asserting objects as the entities on the ontological level, which possess unique well-defined properties. This object-realism could be questioned by Quambiguity, as is done in the approaches of structural realism (Esfeld and Lam 2008; Ladyman 1998a; Worrall 1989).

The variety of possible ontologies in the interpretations of quantum mechanics gives the impression that the ontological level allows arbitrary ontological commitments. This impression is wrong; in fact Kochen-Specker Theorem, Bell inequalities and others ruled out a whole branch of ontologies that are not realised in nature (Kochen and Specker 1975; Zeilinger 2002). This demonstrates, according to the presented theorem on ontological truth, that there is indeed progress in the quest for something like ontological truth nonetheless there might be no final answer to it. This will be discussed in Sect. 5.5

In any case, we want to conclude that Ockham's razor cannot solve the problem of underdetermination and the several interpretations in quantum mechanics in a swift way. It is a concept of parsimony of thought, but was not meant to give a guideline that nature has to be simple in principle. It originates in the economy of thought that is an implication of the limited resources of our cognitive skills, which reflects in well-known heuristics in social psychology. For Ockham himself, the principle was a heuristic of common sense that helped him in his line of argument against Platonism, against a realist position in the debate on the universals. He used the razor to get rid of a not necessarily needed multiplicity of entities.

In context with quantum mechanics, Ockham's razor should encourage us to reconsider the principles and mechanism of our thinking and reasoning, our judging about certain theories, world-views and mindsets, and our prejudices.

5.5 Statements About the Ontological Level

Statements that refer to metaphysical entities refer to the Level 4 of the classification scheme presented previously. Logical Empiricism denied the existence of statements of that kind. We present our views on possible statements on the metaphysical level.

5.5.1 Ontological Coherence

The ontological coherence argument (OCA) is a heuristic that supports the process of theory development from the perspective of ontological consistency in a network of physical theories. Those theories offering ontological elements that fit coherently with ontological elements from a different regime are preferable. By the mutual consolidation of the ontological implications, the OCA supports a network of theories that provides coherent ontologies. Thereby, OCA strongly supports scientific realism and related ideas.

Theorem 5.1 *Under the assumption of a set of Theories $T_{PR\mu}$, which covers different regimes of physics, and that $T_{PR\mu}$ are theories in agreement with empirical data in the respective regime, the OCA claims that those theories $T_{PR\nu} \in T_{PR\mu}$, whose ontologies fit coherently together, are preferable.*

We define a set of elements of a theory T_{PR} and a set of elements of reality PR:

$$T_{PR\nu} = \{t_{\nu i}\} = \{A_\nu, B_\nu, C_\nu, ...\} \tag{5.8}$$

$$PR_\nu = \{\epsilon_{\nu i}\} = \{\epsilon_{\nu A}, \epsilon_{\nu B}, \epsilon_{\nu C}, ...\} \tag{5.9}$$

The set of elements of the theory and the elements of reality obey correspondence rules that are part of the theory. We do not put detailed emphasis on these correspondence rules. A detailed account on the mathematical definition of these rules is provided by Ludwig and Thurler (2007).

We call two theories T_{PR} theory-coherent if their corresponding elements of reality fulfill the condition

$$PR_\alpha \cap PR_\beta \neq \emptyset. \tag{5.10}$$

This implies non-empty sets of elements of reality shared by two theories. In the framework of scientific realism and its related ideas, this provides an argument for an ontological value of those elements of reality (Figs. 5.6 and 5.7).

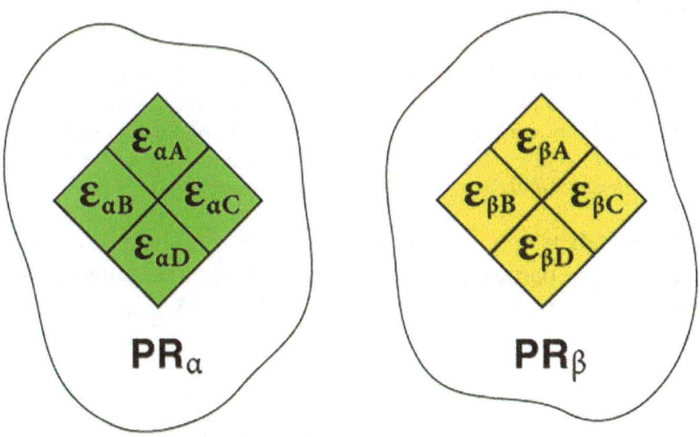

Fig. 5.6 The sets of elements of reality of two theories

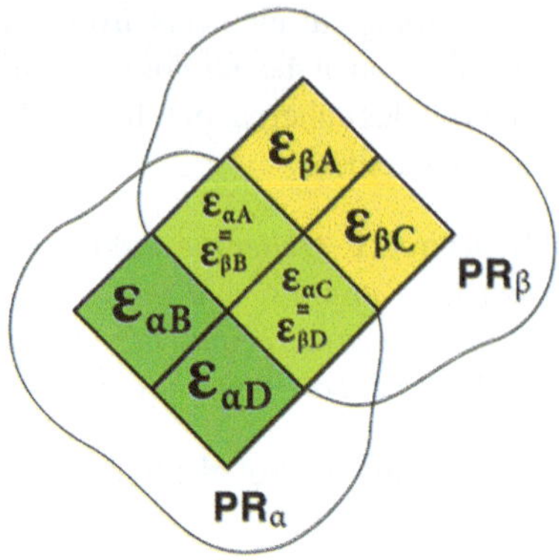

Fig. 5.7 Two sets of elements of reality of two theories

The idea of OCA is related to Ludwig's account on the structure of scientific theories. He introduced the reality domain of a physical theory and stated on the constructed reality domain of all physical theories (Ludwig and Thurler 2007, p. 14):

> *We can now add that the reality domain W is the domain of all Ws, i.e., the Ws of all PTs. Given that all PTs are not known, W cannot be established. By finding new PTs, one discovers new Ws(e.g., atoms and elementary particles). The physically recordable domain W remains decisively limited by the fact that one does not permit all directly ascertainable facts such as, e.g., that a sound is harmonious or that a violin has a good sound.*

And in an earlier account (Ludwig 1978, p. 186):

> *Durch die Weiterentwicklung der Physik wird der durch alle PT's zusammen erfassbare Teil der Wirklichkeit immer größer. Beim Aufstellen des Wirklichkeitsbereichs W einer PT kann bei diesem Entwicklungsprozess des Erfassens eines Ausschnittes der Wirklichkeit kein Widerspruch auftreten...; dagegen sind beim Zusammenfügen der Wirklichkeitsbereiche der verschiedenen PT's im Laufe der historischen Entwicklung der Physik immer wieder Widersprüche der einzelnen Wirklichkeitsbereiche untereinander und der Aussagen der Theorie mit der Erfahrung entstanden.*

> *By the evolution of physics the determined part of reality is growing. In the definition of a reality domain W of a physical theory PT no contradiction can arise in this process of recognition of a part of reality...; In contrast in the history of physics during the merging of reality domains of different PT's contradictions arose at all time within reality domains and between statements of theories and empirical evidences.*[19]

One specific contradiction in the reality domains that Ludwig refers to is the contradiction between the particle picture and the wave picture in quantum mechanics. We provide an account of this contradiction in Sect. 5.5.2, after the definition of ontological truth.

---

[19] Translation by the author.

5.5.2 *Ontological Truth*

Logical Empiricism avoids to provide ontological statements. In our context of the classification scheme, this alludes to statements that refer to the Level 4. Statements of that kind enclose metaphysical content and are superfluous in the sense of Logical Empiricism's doctrine to confine itself only to statements on experience. If we ask for truth values of statements on metaphysical content, Logical Empiricism would deny that such statements are meaningful.

Critical Rationalism would also deny that truth values on metaphysical statements are meaningful at all. In Critical Rationalism, the argument would be that metaphysical statements cannot be falsified by experience if the metaphysical content does not refer to any experience (Popper 2005).

We claim that this does not imply that scientific knowledge is not growing, or that we can not learn more about how the world "really" is. Assuming that it is possible to learn about such a reality, which exists independent of us, is the basic assumption of scientific realism and all its related positions. The assumption is that there are statements on the metaphysical level that have truth values.

Objections against such a view are manifold. Most important to the author are Scientific Underdetermination of theories (2016) and the Pessimistic Meta Induction (PMI) (Laudan 1981). A summary of the PMI argument is provided by Ladyman (2016):

> *Proposition p is widely believed by most contemporary experts, but p is like many other hypotheses that were widely believed by experts in the past and are disbelieved by most contemporary experts. We have as much reason to expect p to befall their fate as not, therefore we should at least suspend judgement about p if not actively disbelieve it.*

For recent debates on the PMI argument, see Fahrbach (2009), Park (2011), Müller (2015).

In favour of scientific realism, Hilary Putnam formulated the Miracle argument (Putnam 1975), which basically states that the success of explanation and prediction of our best scientific theories would be a miracle if these theories do not refer to a given reality beneath the level of theories.

Mathematical realists would see here a way with their approach of ontological reality of mathematical entities to solve the "miracle" in the miracle argument, the unreasonable effectiveness of mathematics (Wigner 1960) and the question on the nature of this metaphysical reality in one stroke.

The question on the nature of reality is still open, and as the recent discussions on structural realism demonstrate, an end is not in sight. Too many questions are open or unclear and need a thorough analysis, but we want to present a simple argument that provides a preliminary and confined answer to the question of scientific knowledge and scientific progress:

Theorem 5.2 *There exist statements on truth values on the level of ontologies that result from empirical science. These statements always are exclusive statements on ontologies.*

Example: The earth is not a disk.

Example: The earth is not a sphere.

These statements are both true and can claim to refer to the ontological level. We can state that the earth is not a disk, since we have an idea about the theoretical term "disk" and its ontological meaning. From experience, we could deduce that the earth is not a disk. And the same for the statement that the earth is not a sphere. The earth has an eccentricity and has properties that definitely do not resemble those of a sphere. That is a statement that follows from experience and refers to the ontological level of the entity earth.

We cannot provide an answer as to what the earth or any ontological entity is, but we have access to empirical statements that help us in our quest on scientific knowledge. One could give the objection that this is not very satisfying, because the realm of those metaphysical entities which the earth potentially resembles is very large, probably infinitely large. The statements we can provide are finite, giving the impression that although we gain scientific knowledge, we will never reach a definitive end (Fig. 5.8).

Another debate that demonstrates the application of the concept of ontological truth is the discussion on the Wave-particle duality, or in Bohr's conception Wave-particle complementarity. An introduction to the historical development of methodologies of particle and wave-theory of light is provided by Achinstein (1991). An overview on the positions of wave-particle complementarity concerning light is provided by Combourieu and Rauch (1992), Ghose and Home (1996).

The classical concepts of particles and waves are, in principle, untenable in a general way. A particle that is localized at a specific point in spacetime, carrying properties, is outworn in the same way as the concept of a completely delocalized and holistic entity wave that is spread over space and time, reaching into infinity. Both concepts are based on the Level 3 of the classification scheme and represent an idea on an ontological entity. Following the conception of ontological truth, a statement of the kind "light is a particle" or "light is a wave" cannot be a tenable statement on

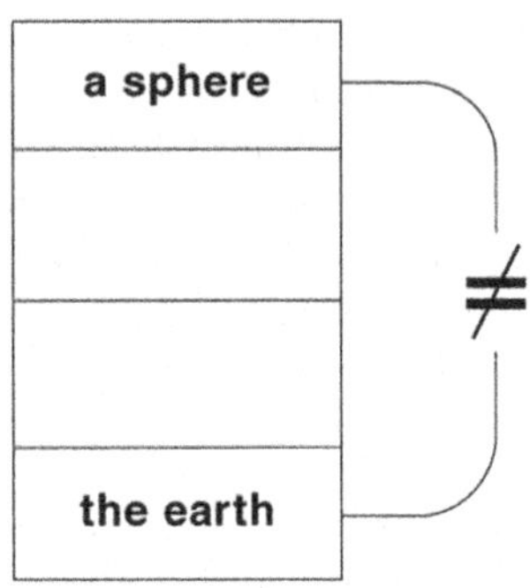

Fig. 5.8 The relation of the terms to the ontological levels in the statement

truth about the ontological entity light. Also, a statement of the kind "light is both particle and wave" is not valid according to the conception of ontological truth.

The negated statements "light is not a particle" and "light is not a wave" would fulfill the requirements on truth statements, as well as "light is neither a particle nor a wave." It turned out that the classical conceptions particle and wave do not represent concepts that fulfill the needs of our best theories (Combourieu and Rauch 1992, p. 1403–p. 1404) provide a review of why these conceptions failed, and Bohr had to introduce a principle to recover the classical conceptions:

> *... no experimentalist has ever been able to show experimentally this "dual" behavior within the same experiment with a single particle and an experimental setup absolutely identical; in other words, to violate the complementarity principle postulated by Niels Bohr in 1927, and since then accepted by the majority of physicists as an operational principle to settle the wave-particle dualism problem. ... No experimentalist ever has succeeded in detecting a single particle travelling in one or the other branch of an interferometer (Michelson, Mach-Zehnder, or pure crystal) while analyzing simultaneously the interference pattern which it is supposed to appear on a screen located just behind the detectors*

Out of our best empirical theories, we get an idea about the properties and behaviour of light; we can model a conception, but never unveil the whole truth about an ontological entity.

Callender (2015) provides an account on resolving the wave-particle duality by embedding the question into a greater philosophical framework and assuming only one type of ontological entity:

> *We have motivated the nomological understanding of the wavefunction. On this picture the It-or-Bit debate, and even the wave-or-particle debate, are absorbed by a larger more philosophical debate over the nature of laws. ...Regardless of whether one explores this option, the main point is that quantum mechanics requires only one type of beable.*

The relation to scientific underdetermination and the unreasonable effectiveness of mathematics should be discussed within the framework of ontological truth and remains an open question.

There is a manifold of philosophical approaches on the question of truth values of scientific statements. One prominent account is provided by Tarski (1944). All those approaches are philosophical and exist somewhat parallel to each other and it seems that they cannot be falsified by empirical evidences. The goal of the stated theorem on ontological truth is to provide an insight on progress in the quest for scientific knowledge, even if that implies to deny or stay agnostic towards a convergent realism. Further, it clarifies that there are statements with truth values even on the ontological level. That these statements have to be formulated in a negative exclusive way resembles the falsificationism in the empirical sciences.

Chapter 6
Summary, Conclusion and Outlook

Quantum mechanics is undoubtedly the most successful theory of physics of the 20th century; it provided a theory for unexplained observed phenomena, it predicted new phenomena, and it provided the technological framework for some of the technological revolutions of the twentieth century. But it also revealed a significant problem with the ontology of physics.

We introduced ontology, or metaphysics, as a philosophical discipline that inquires into the nature and categories of existence. Physics and ontology are interconnected because physical theories often propose an ontological framework alongside their formal structures.

The problem with ontology in physics is often seen as something that emerged with the development of quantum mechanics. But problems with a clear ontology already have been present in classical physics. Newtonian mechanics was highly successful in its formalism, but entailed many problems with the ontology, like action-at-distance. Maxwell's electrodynamics could not explain the structure of matter in the formalism or in the ontology, and the implied conception of fields and particles is pervaded by the conceptual dichotomy of continuity and discreteness. The practical and technological success of classical physics diverted attention from ontological and foundational problems until the advent of quantum mechanics.

With the ascendance of quantum mechanics as a foundational theory, demand in the physical community grew to propose a theory with clear conceptions and ontology. In this environment, debates on the foundations of quantum mechanics emerged, questioning the concept of reality and even the scope and methodology of physics itself. We believe that it is justified to ask for an ontology in physics, and quantum mechanics encourages us to question our methodology to provide ontologies, or question if we have reached a point where we cannot speak about a clear ontology at all. These questions touch our notion of reality; therefore, we have to define realism in the first place.

Realism concerns metaphysics, addressing what exists and how our best theories, those with the highest explanatory power, relate to the nature of reality. To understand what metaphysics means, we have to distinguish it from physics. Physics, as a set of methodologies, is based on our perceptions, on phenomena. One methodology

G. C. Krizek, *The Conceptions of Reality in the Interpretations of Quantum Mechanics*,
https://doi.org/10.1007/978-3-031-65786-3_6

of physics is the application of quantitative methods to support our thinking about phenomena.

Our thinking about our perceptions, about the phenomena, can be guided by rationality or empiricism. Either the theories we produce are a description of a metaphysical world, or our theories are models that do not have any connection to the phenomena itself or their causation, besides their effort of our thinking. The first position would imply metaphysical realism; the latter one is an anti-realist position. In the debates on quantum mechanics, anti-realist positions claim that an ontology is not needed; physics is a pure empirical theory that does not need the attachment of an ontology.

In this tension, the development of quantum mechanics challenged the conceptions of reality as much as the theory of relativity did for the ontology of space and time. It sparked a discussion that has not come to an end even now. Specifically, the discussion on the EPR-paradox and the Bell inequality demonstrated how deeply related physics and philosophy are, since with the Bell inequality physics ruled out the whole philosophical conception of local realism in a rather convincing way.

Since the interpretation of quantum mechanics challenged the methodology of physics, it has become a part of some interpretations of quantum mechanics to reconsider the purpose and scope of physics as a whole. Therefore, it makes sense to shed some light on the relation of physics and philosophy.

Physics, on one hand, is a methodological confined continuation of philosophy of nature, a specific branch in the broader enterprise of science. Philosophy, on the other hand, advances a freedom in the choice of methods. There is no methodological restriction. Physics is methodologically defined by philosophy of science. There is a certain dimissive tradition in parts of the physics community concerning the relevance of philosophy to physics, which manifests in the rejection of ontological discussions and the limitation of physics to empiricism. We do not follow this tradition since we argue that rejecting the methods of philosophy is a rejection of the methodology of physics, since the methodology of physics is justified by philosophy of science.

On the other hand, in philosophy there might be an underestimation of physics as methodology to give insights into the very nature of reality. Physics provides conceptions that are successful in quantitative prediction, helpful in understanding and explaining phenomena, and enlightening in the prediction of new, previously unobserved phenomena. The predictive power of physics encourages proponents of metaphysical realism and related positions to believe that the theory contains some true proposition about the nature of reality, since something new and unobserved has been revealed just out of rationality. The miracle argument is based on this observation. Out of these observations, proponents of realism draw conclusions on ontological entities and their relation.

In that sense, physics and philosophy are entangled in their evolution and methodologies, and it is worthwhile to consider both fields and their respective state of knowledge to overcome problems that arise in the other field.

One methodological approach of philosophy of science we discussed in this thesis is Ockham's razor. This famous heuristic found its way into the debates on the interpretations of quantum mechanics as an argument against certain interpretations.

We presented the origins of Ockham's razor and the connection to the problem of universals, which was a current problem in medieval philosophy. The problem of universals deals with the question of whether the categories exist in nature or if only singular entities exist. To put it differently, whether the colour "red" has an existence in an object on its own, or if the redness of the object is a categorisation of the observer and his thinking. The first position is realism, and the latter one is the position of nominalism. Ockham proposed nominalism as an answer to the problem of universals, and his line of argumentation was that entities that turn out to be unnecessary should be abandoned, because they can be multiplied in an arbitrary way. Therefore, he rejects ontological relevance to these universals.

Ockham's razor has not been formulated as a principle by its founder; it has been applied on several occasions in his writings and pervades his philosophy. Also, it is not seen as a principle of nature in the sense that nature has to be simple, or that only simple theories are the correct ones. To Ockham, it is a common sense principle related to economy of thought, and not related to nature in the first place. Based on a work by Thorburn (1918), we presented passages in translation in Ockham's work where the razor principle is applied.

In quantum mechanics, Ockham's razor became an attractive and often used heuristic to attack certain interpretations, due to the simplicity argument. By the methodology of comparison, we contrasted the application of Ockham's razor between different interpretations of quantum mechanics, and it turned out that proponents of all interpretations apply Ockham's razor against each other with the same argument, that their proposed interpretation is more simple that the other one. This balance of refusal demonstrates the subjective character of simplicity in the discussion on interpretations of quantum mechanics. Even if a measure for simplicity would be defined to get rid of the subjective account, the debate would probably shift on the definition of this measure that would claim to give an insight on ontological truth by its simplicity. Therefore, we proposed that Ockham's razor as a heuristic cannot contribute to the debate in the interpretations of quantum mechanics. Furthermore, we demonstrated that Ockham's razor should not be understood as a principle of nature.

In quantum mechanics, we notice a diversity of interpretations that overlap in their empirical predictions but disagree in their ontological conceptions. Out of this, we concluded a resemblance to the Great Cosmological Controversy, where the ambiguity between a geocentric and heliocentric world-view (although we decided in favour of the heliocentric world-view) turned out to have a certain reason in the principle of relativity.

We concluded that the current situation in quantum mechanics could be subject to an analogy to the Great Cosmological Controversy in the sense that the ambiguity we recognise in the interpretations of quantum mechanics has foundational origins that go beyond our actual perception. We called this feature of quantum mechanics "Quambiguity," reflecting the many ontological aspects that are involved in quantum mechanics. In any case, we argue that it is not the time to halt discussions on interpretations of quantum mechanics based on heuristics like Ockham's razor.

Quambiguity suggests that the various interpretations of quantum mechanics are all valid for a reason, representing different aspects or facets of a larger, underlying reality. If that is not the case, the conundrum of the interpretations of quantum mechanics would lie somewhere else. To give an overview on the interpretations of quantum mechanics and the different aspects of the formalism, their correspondence, and their contradictions and differences, we proposed a classification scheme. The proposed partially hierarchical scheme consists of four levels to distinguish between the mathematical formalism, the assignment of mathematical objects to names and terms, the concepts in the theory, and the related ontological commitments.

In Chap. 2 we defined the levels of the classification scheme in the following way:

Definition. Level 1 of a physical theory is described by a set of mathematical quantities. These mathematical quantities are in relation to each other by mathematical laws.

Definition. Level 2 of a physical theory is described by a set of measurement mappings to the mathematical quantities of Level 1. It describes how the defined quantities in the mathematical description of the theory are related to sensations and experience. It relates mathematical objects to theoretical terms.

Definition. Level 3 of a physical theory is described by a number of concepts and principles that are related to the mathematical description of the theory. These principles can be formulated as statements or logical expressions.

Definition. Level 4 of a physical theory contains the ontology of the theory and the metaphysical entities or conceptions that are introduced by the theory.

In a figurative way, the different levels of the classification scheme can be represented by the following symbols. A dot represents a mathematical symbol, which is connected to other mathematical symbols by a mathematical relation. This mathematical relation is depicted by a line between the involved symbols. A concept is symbolised by a hatched area confined by the used symbols and laws. An element of objective reality is depicted by a rhomb.

On Level 1 of the classification scheme, there are only the mathematical symbols and their connecting laws. On Level 2, the mathematical symbols receive names and are assigned to empirical quantities. Thereby the mathematical formalism becomes a physical law. On Level 3, the concepts and principles that are used in the theory, which involve the physical quantities, are defined. Level 4 refers to the elements of objective reality, the entities of an ontology of the specific theory (Tables 6.1 and 6.2).

For a graphical representation of these elements, refer to the following table:

These graphical elements can be applied to the four levels of the classification scheme to represent physical theories.

A physical theory is then represented by the following symbol (Fig. 6.1):

We applied this classification scheme to classical concepts of reality used in the EPR argument; specifically, the concept of naive realism, elements of reality and the

Table 6.1 Figurative elements used in the classification scheme

	Physical quantity
	Law between physical quantities
	Concept connecting laws and physical quantities
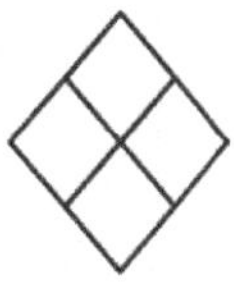	Elements of objective reality

criterion of reality. We presented a definition of naive realism and demonstrated that Einstein was not a proponent of a naive realism, although he had a kind of mathematical realism in mind. We clarified some ambiguous applications of elements of reality in the EPR paper, and showed that elements of reality should not be understood in the sense of a naive realism, but more in the sense of an ontological entity that corresponds to quantities or concepts in the theory. An element of reality itself is only comprehensible through sensations, although it is thought to exist in the objective reality. The ontological entity itself is not accessible, but corresponds to quantities or concepts in the theory.

The criterion of reality has been discussed in the framework of the classification scheme, and we presented its surprising mechanism that suggests elements of reality out of the theory. This is done by the methodology of identifying elements in the theory that obey probability one, and conclude that they have corresponding elements of reality with ontological character. If we apply this methodology to classical physics that fulfills determinism, all physical quantities obey a probability measure of one. Therefore, in classical physics, all physical quantities should have a corresponding element of reality, which is not the case, as we demonstrated by means of Newtonian mechanics. The principle of invariance supports this process of identification of ontologically relevant quantities and concepts in the theory. In the EPR argument, this invariance plays no key role, and the chosen quantities are assumed to fulfill it implicitly.

Furthermore, we demonstrated that against the claim of Fine, the criterion of reality has been applied in a letter to Popper. Therefore, it is conclusive to assume that Einstein was aware of the criterion of reality, that it was not only Podolsky's

Table 6.2 Classification scheme

1	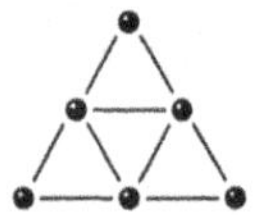	Mathematical laws between mathematical symbols
2	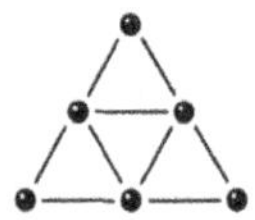	Interpretation physical quantities Relation physical quantities - Measurements Measurement laws
3		Concepts and principles
4		Ontology - beables - elements of reality

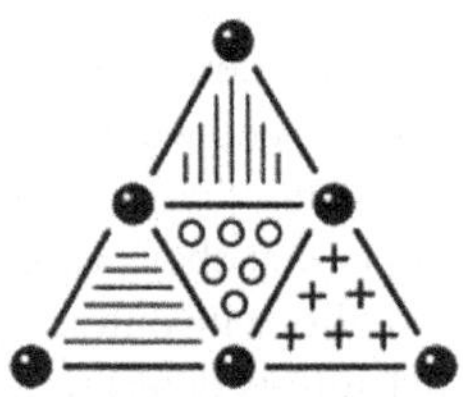

Fig. 6.1 A physical theory in the classification scheme

proposition, and that he advanced it until he found an easier way to formulate the EPR argument. A comparison of five versions of the EPR argument is subject to future work.

We presented cornerstones of Einstein's conception of realism, since he never provided a consistent scheme. His ideas involve objectivity or a kind of metaphysical realism, mathematical simplicity as something realised in nature, a four-dimensional spacetime continuum as the foundational entity, methodological realism, phenomenal realism, deterministic character, local causality, unambiguousness, reductionism and singularity-freeness and nonlocalizability. These conceptions are to some extent involved implicitly or explicitly in the EPR argument.

The purpose of the EPR argument to Einstein can be seen in a different light when the Einstein-Rosen paper is taken into consideration. One core aspect of the debates on the interpretations of quantum mechanics is the concept of completeness. After several unsuccessful attempts previously, Einstein challenged the completeness of quantum mechanics in the EPR paper in a way that influenced and shaped the debate up to now. We proposed that there is a connection between the EPR paper and the Einstein-Rosen bridge paper of the same year, since both papers ask for completeness of a theory. The ER paper is famous for presenting one of the first wormhole conceptions in general theory of relativity. It is tempting to think that Einstein had in mind solving the nonlocality issues of quantum mechanics by wormhole solutions of the metric field. We demonstrated that for several reasons this idea is unjustified, but there is still a deep connection between those two papers, which refer to the core idea of the EPR argument, the concept of completeness.

We provided definitions for several versions of completeness, which differ in their reference to a level in the classification scheme, but also in subtleties in their general definition.

Definition. A theory fulfills theory completeness if its physical quantities and laws are connected by concepts that relate only to each other, and the theory renounces extra assumptions of concepts that exceed the framework of the theory.

Theory completeness refers to the Levels 1–3 of the classification scheme, not to an ontology. It only refers to elements of the theory, like physical quantities, laws or concepts, and requests an inner consistency of the theory.

Definition. Bijective completeness means that every element of the physical reality ϵ_i must have a counterpart t_i in the physical theory.

Though *bijective completeness* is a Bijection between elements of the theory and elements of reality, it starts with the latter ones. Starting at elements of reality causes the problem of how of to identify them and the corresponding elements of the theory. For this identification, Einstein defined the completeness criterion.

Definition. Born completeness for a theory is fulfilled if the theory contains no probabilistic elements. A theory which contains statistical accounts and probabilistic statements is considered to be incomplete.

By *Born completeness*, quantum mechanics is incomplete per definition. No further arguments are needed.

Definition. A theory containing probabilistic statements about elements of the theory is Schrödinger complete if those statements refer to single events or entities and are fundamental in the sense that nothing more can be said about the system. The theory is considered to be incomplete if there are potential new elements in the theory that would abandon the probabilistic character of the theory.

A theory containing probabilistic statements about elements of the theory is Schrödinger complete if those statements refer to single events or entities and are fundamental in the sense that nothing more can be said about the system.

The theory is considered to be incomplete if there are potential new elements in the theory that would abandon the probabilistic character of the theory.

The definition of *Schrödinger completeness* refers to what we would nowadays call "hidden variables." The probabilistic character of the theory is assigned to single events. A theory is incomplete in the sense of *Schrödinger completeness* if there is an underlying mechanism to explain the probabilistic character of the single events. If a theory is incomplete in the sense of *Schrödinger completeness* because of the lack of an underlying mechanism, it is also incomplete in the sense of theory completeness.

In the reading of the Brukner-Zeilinger interpretation, quantum mechanics is complete in the sense of *Schrödinger completeness* and in the sense of theory completeness, since there is the concept of irreducible randomness, which provides an explanation for the probabilistic character of the single events on the conception level of the classification scheme, and the theory is internally consistent.

We amended the presentation of completeness conceptions by the definition of *standard completeness* by Held and *ψ-completeness* defined in Harrigan and Spekkens (2010):

> *An ontological model is ψ-complete if the ontic state space Λ is isomorphic to the projective Hilbert space $\mathcal{PH}$ (the space of rays of Hilbert space) and if every preparation procedure P_ψ associated in quantum theory with a given ray ψ is associated in the ontological model with a Dirac delta function centered at the ontic state Λ_Ψ that is isomorphic to Ψ, $p(\lambda|P_\psi) = \delta(\lambda - \lambda_\psi)$.*

The concept of *ψ-completeness* is setting up an Isomorphism between the projective Hilbert space and an ontic state space, whereas *bijective completeness* is a Bijection between elements of a theory and elements of reality. These elements of reality have no defined mathematical structure, per se. Since *bijective completeness* refers to an assignment of elements of the theory to elements of reality (in the language of the classification scheme, some assignment of elements of the theory Level 1–3 to the ontological Level 4), the concepts of *bijective completeness* and *ψ-completeness* do not refer to the same structures, and therefore, cannot be seen as identical. With these definition of completeness, we analysed the EPR argument in the light of the connection to the ER paper.

We proposed that Einstein's intention of writing the EPR paper was to demonstrate the incompleteness of quantum mechanics in the sense of *Schrödinger completeness*, to justify his ambitions to look for a complete theory in the sense of *theory completeness*. With this new theory programme, which he presented in the ER paper in outlines, he aimed to explain the atomistic character of quantum mechanics and reconstruct quantum mechanics out of field theory. We identified that "atomistic character" refers to the singularity-free description of particles, which is the core topic of the ER paper.

The argument for the incompleteness of quantum mechanics presented in the EPR paper was a justification for a research programme that was more important to

Einstein, the programme that later developed into the unified field theory programme, on which he worked for the rest of his life.

By applying the separation principle in the EPR argument and thereby excluding the measurement process from the consideration, Einstein aimed to show that the statistical character of quantum mechanics is not owed to the measurement process, but resides only in the state vector formalism and the Schrödinger evolution. Since the state vector formalism and the Schrödinger evolution are fully deterministic themselves, he concludes that the formalism of quantum mechanics has to be incomplete in the sense of *Schrödinger completeness*.

The separation principle had the purpose to confine the discussion on the completeness of quantum mechanics to the formalism and the Schrödinger evolution, and exclude the difficult concept of measurement. Since the EPR argument was a complicated construct with some assumptions, it was easier for Einstein to justify his aim for a new theory with a simpler argument. He challenged *Born completeness*, and thereby the whole statistical character of quantum mechanics, and he shifted the effort from a critique of quantum mechanics to development of a unified field theory.

The idea that Einstein might have aimed for an explanation of nonlocal phenomena in quantum mechanics by Einstein-Rosen bridges has no basis at all, since to Einstein, in quantum mechanics there was no nonlocality issues, and the purpose of the Einstein-Rosen bridge solution was to get rid of the localizability of particles in a field theory. This tempting idea reflects in recent claims by Maldacena and Susskind (2013), Susskind (2016), which propose an EPR = ER principle.

One further aspect of the classification scheme that has been presented is its connection to related concepts on the structure of scientific theories in philosophy of science. We presented several systems and classifications for physical theories that developed in the 20th century, starting from Einstein's model of layers, to the received view, the semantic view, and the structuralist approaches up to structural realism. The latter one turned out to be an interesting stance, in context with quantum mechanics, and has been presented in more detail.

Structural realism was developed to credit the critique of The Pessimistic Meta Induction and the Miracle argument. It is the attempt to reconcile these two strong philosophical arguments, where the Pessimistic Meta Induction points in the direction of anti-realism, whereas the Miracle argument advances a realist position.

Structural realism is divided into three positions, epistemic structural realism, ontic structural realism and moderate structural realism. In general, all positions of structural realism share the emphasis of structures as the primary entities of a theory as well as the ontology. The status of the objects, as the relata in the structural relationships, is seen differently in the three approaches.

Epistemic structural realism advances the position that there exists a reality for objects, but not in an accessible way. Ontic structural realism abandons the idea of objects, and specifically individual objects as necessary to more than being placeholders in the structures. Moderate structural realism criticized this stance of ontic structural realism by stating that the relata, the objects, in the relations are still needed. Moderate structural realism emphasizes that structures and objects are on the same ontological footing.

In context with the ontological level of the classification scheme, we proposed two conceptions, the Ontological Coherence Argument and the conception of ontological truth.

The Ontological Coherence Argument has been presented in context with Ockham's razor and other heuristics that influence ontological commitments. It provides a guiding principle for theory selection, if the related theory and its ontological commitments can be put in relation to other theories and their ontological statements. It is then desirable to adopt those theories that are coherent in their respective ontologies. This heuristic is based on the assumption of metaphysical realism.

Theorem *Under the assumption of a set of theories $T_{PR\mu}$, which covers different regimes of physics, and that $T_{PR\mu}$ are theories in agreement with empirical data in the respective regime, the Ontological Coherence argument claims that those theories $T_{PR\nu} \in T_{PR\mu}$, whose ontologies fit coherently together, are preferable.*

On ontological truth, we proposed a theorem that reflects the scientific progress and states that scientific knowledge is cumulative but not convergent.

Theorem *There exist statements on truth values on the level of ontologies that result from empirical science. These statements always are exclusive statements on ontologies.*

The exclusive statements that can be justified by our best scientific theories in the sense of the ontological truth theorem are relations between an element of reality and an element of the theory. By stating what the world is not like, we learn more about the nature of reality, but not in a way that will unveil and narrow down reality as a whole with a number of finite statements.

To demonstrate the effectiveness of the classification scheme, we applied it to classical mechanics, Bohmian mechanics and the De Broglie-Bohm interpretation. By comparison of the involved concepts, the used formalism and the ontology, we demonstrated that Bohmian mechanics differs vastly from the De Broglie-Bohm interpretation. The mathematical formalism is nearly identical, and the measurement mappings are the same, but the concepts in the theory differ from each other. In the concepts of reality, both interpretations completely disagree. Though Bohmian mechanics emphasizes the particle ontology as foundational, the De Broglie-Bohm interpretation emphasizes a novel matter ontology, which goes beyond just assigning particle and wave properties to the quantum system. It represents the view that matter behaves in a novel way in the quantum regime that is incompatible with both ontological pictures of particles and waves alike. The matter in the De Broglie-Bohm interpretation is something different.

To give a better idea about the realism concepts, we provided an overview on realism in classical physics, metaphysical realism, scientific realism and object ontology, and completed the presentation with conceptions of reality in quantum mechanics. These involve naive realism, the EPR reality criterion, Ruark's criterion of reality, Bell's local realism, and realism conceptions in context with the Kochen-Specker theorem.

In all those conceptions, the notion of an object plays a significant role. We defined an object as an entity capable of possessing a set of properties that distinguishes it from other objects and its environment. Quantum mechanics challenges this conventional understanding of objects.

Mathematics over the real numbers provides a methodology to assign real numbers to a mathematical symbol as a representation of properties of an object, like the quantity of mass of a rigid body. The formalism of quantum mechanics offers something completely new. With the linearity of states on the Hilbert space and the tensor product, a new conception seems represented. Properties do not belong to an object alone anymore; they can be shared among several objects. This feature, entanglement, is one of the essentials of quantum mechanics. It reflects what is emphasized by structural realism, that structures in the sense of relations between relata are the relevant entities.

Esfeld (2013) emphasized that structural realism alone cannot solve the conundrums of quantum mechanics, since structural realism can give no answer on how the structures are instantiated. This question has to be answered by the theory and its interpretation.

A presentation of how three major interpretations instantiate the structures has been provided by Esfeld (2013), but he omitted several other approaches in the interpretations of quantum mechanics. We amend this presentation by three interpretational approaches; the De Broglie-Bohm interpretation, the Brukner-Zeilinger interpretation and QBism. These three approaches share their emphasis for informational concepts, but in completely different directions of in-formation, information and belief.

In the De Broglie-Bohm interpretation, the Quantum potential is the mathematical object to instantiate the structure. It involves all particle positions as relata in a natural way, directly out of the reinterpretation of the Schrödinger equation. For k particles, the Quantum potential is defined as

$$Q = -\sum_{k=1}^{N} \frac{\hbar^2 \nabla_k^2 R}{2m_k R}.$$

The real-valued function R depends in general on the position of all k particles:

$$R = R(\vec{q}_1, ..., \vec{q}_k, t)$$

Proponents of the De Broglie-Bohm interpretation emphasize the Quantum potential as an information potential, but with a different conception of information, in the sense of in-formation referring to Aristotle's formal cause.

In context with structural realism the De Broglie-Bohm interpretation identifies active information and its mathematical carrier, the Quantum potential, as the resource representing the structure that is requested by a structural realist approach. In contrast to Bohmian mechanics the De Broglie-Bohm interpretation puts no big emphasis on the concept of the particle. The emphasis lies in the structure of the theory, which represents the holistic character of nature and the impossibility to describe nature in a correct way with reductionist approaches.

The Brukner-Zeilinger interpretation is a reconstruction of quantum mechanics based on informational principles as foundation. It utilizes the conception of a hypothetical observer to omit the assignment of the quantum state to a specific observer. Information is the foundational entity on the ontological level, and quantization follows from of the limitation of what can be said about the world. Reality and facts about the world are defined in a relational way, which is supported by structural realism.

The Brukner-Zeilinger interpretation remains neutral on the ontological structure of the objects on the quantum level, but, on the other hand, it clearly abandons the idea of classical pre-existing properties, besides the informational content of the elementary systems.

In the Brukner-Zeilinger interpretation, the informational storage capacity of an elementary system and its contained total information is identified as structure, in the sense of structural realism. Entanglement, then, is the practical feature of quantum systems that shares a structure of common or joint information.

QBism is a reconstruction of quantum mechanics that emphasizes the subjective character of quantum mechanics, or merely the mixture of subjective and objective elements in one formalism. The mixture of this is subjective, and objective elements are the source of the difficulties with the interpretation of the formalism of quantum mechanics.

A tenet of QBism is the subjectivity of probabilities. They represent only the individual beliefs of an agent on some specific event. The quantum states thus have no ontological footing in the world.

Relating QBism to structural realism, but neither ontic structural realism nor epististemic structural realism, seems suitable to represent the position of QBism. Both structural realist positions do not consider the conception of beliefs of an agent, as emphasized by QBism. Therefore, we proposed QBism as a different form of structural realism, a doxastic structural realism.

The structures that are instantiated by QBism are the mutual beliefs of agents performing quantum mechanics. These structures represent an internalist view of the beliefs on the behaviour of nature of one specific agent in interaction with nature. If we further refer to the old methodology of philosophy and physics to search for invariants in the models, in order to identify ontological relevant quantities, QBism proposes Hilbert space dimension as a invariant capacity, which is unchanged for a specific quantum system. In the sense of structural realism, the Hilbert space dimension is a structure with objective character. Results from particle physics on the neutral K-meson system support the subtle character of the Hilbert space dimension (Di Domenico et al. 2012; Hiesmayr 2007). The decay process of the neutral K-meson for instance does not involve an extra Hilbert space dimension, though it is an effective two-state system.

These three approaches to the interpretations of quantum mechanics share their emphasis on information, but provide fundamentally different conceptions for this term, although they agree in some points. QBism and the De Broglie-Bohm interpretation agree in the observation that a quantum state, as described by the state vector, is a mixture of elements that do not really belong together. They disagree on the nature and structure of these elements. To QBism, it is a mixture of subjective and objective elements, whereas the De Broglie-Bohm interpretation decomposes the Schrödinger evolution into an extended Hamilton Jacobi formalism and a continuity equation for probabilities.

QBism and the Brukner-Zeilinger interpretation agree in the emphasis of the significance of the Hilbert space dimension, but disagree on the interpretation of the very nature of it. Brukner and Zeilinger assign the conception of information to the elementary system, whereas QBism speaks more generally of a capacity of the quantum system.

QBism and Brukner-Zeilinger interpretation agree on the nature of information as something related to knowledge, perception or belief of observers, although they disagree on the question of whether this information is objective or subjective.

All of them agree that the conception of objects is not compatible with quantum mechanics and its interpretations. In the context of Bohmian mechanics, an observation has been made on space as degree of freedom, with an exceptional position under all degrees of freedom in quantum mechanics. In the framework of Bohmian mechanics, in the reading of moderate structural realism, the position of a particle, in the sense of Bohmian mechanics, remains the only intrinsic property left. All other properties turn out to be contextual and support the idea of abandonment of the object-character of quantum systems. The position plays a distinguished role in Bohmian mechanics; this distinguished role reflects in the formalism of quantum mechanics as the position is assigned to an infinite-dimensional Hilbert space. By these observations, the concept of an object itself becomes dubious and seems to lose its ontological footing.

The ontological conception of an object was highly successful in classical physics and compatible with our common sense experiences; thus, it is deeply anchored in our thinking and our scientific theories. But even though it is so well-founded in our common sense experience, the ontology of an object is not a success story. All entities that have been proposed to be objects turned out to be constructed out of deeper building blocks. Physics never came to the point of really facing an object at all. Atomism showed us the road to these fundamental building blocks of matter, and quantum mechanics set the rules for their behaviour, but we have no good reason to believe that these fundamental building blocks are objects on the conceptual and ontological level.

Structural realism might only be the start to understanding, that in the quantum regime, the world is not understandable with the help of reductionism, and probably holism can provide a framework that represents the nature of reality in a more suitable way. The challenge for physics, then, is to adapt its formalism and methodologies to an enterprise in a holistic universe where we obtain an internalist perspective.

Structural realism on one hand complies with the miracle argument of scientific realism, and on the other hand considers the critique of Pessimistic Meta Induction, in the sense that our best theories, those with the highest explanatory power and empirical confirmation, always speak about structure, and it is structure that reflects the ontological elements. By this change of perspective on structures as the primary ontological relevant entities, structural realism resolves the critique of Pessimistic Meta Induction.

The formalism of quantum mechanics and its difficult interpretation suggests a modification of our philosophical conceptions about reality in several possible ways. One conception that we might have to give up is the ontological conception of objects. Structural realism suggests that what is out there is not an object-like entity that carries properties; it suggests that what is there is a structure that can be revealed by our best scientific theories. Entanglement turns out to be a key feature that reflects this structural realism in quantum mechanics. Entangled quantum systems do not possess properties in the individuals; what can be assigned, instead, are relations between the entangled quantum systems.

The plurality of approaches in the interpretations of quantum mechanics can be seen as annoying or enriching. That it is possible for a theory to provide such a variety of meaningful interpretations is surprising, and we want to encourage the physics community to learn from this variety instead of dogmatically fighting it.

The De Broglie-Bohm interpretation demonstrated that it is possible to give an interpretation of quantum mechanics based on hidden variables, with holism as ontological perspective. The Many worlds interpretation demonstrated how an interpretation of the formalism of quantum mechanics without amendments in the formalism can provide an ontology. Bohmian mechanics demonstrated the significance of position as a distinguished property and hidden variable of a quantum system. The Brukner-Zeilinger interpretation demonstrated how quantum mechanics can be seen as a theory about information of hypothetical observers of quantum systems. QBism showed that it is possible to give a continuation of the Copenhagen interpretation with a subjective perspective that retains realism.

All those interpretations of quantum mechanics provided insights on aspects of nature in the microscopical regime. None should be abandoned. It remains an open question as to how it is possible in the framework of structural realism that these structures lead to such a manifold of feasible interpretations that are coherent and overlapping in their formalism and empirical predictions, but orthogonal in their ontological commitments. Is this characteristic an indication for something lying underneath that points in the direction of Quambiguity, to justify the conundrum of these manifold interpretations and their orthogonal ontologies? Or is the ambiguousness just owed to scientific underdetermination?

If nature cannot be seen in an unambiguous way on this microscopical scale, it should not surprise us at all. Our logic is based on common sense, inspired by observations in the classical regime which built the basis of Aristotelian logic. In any case, physics should lead us to adjust our philosophical perspectives based on empirical observations and metaphysical insights, rather than the other way around.

Appendix

G. C. Krizek, *The Conceptions of Reality in the Interpretations of Quantum Mechanics*,
https://doi.org/10.1007/978-3-031-65786-3

A.1 Appendix—The EPR Paper—Refining the Notions of Completeness

of lanthanum is 7/2, hence the nuclear magnetic moment as determined by this analysis is 2.5 nuclear magnetons. This is in fair agreement with the value 2.8 nuclear magnetons determined from La III hyperfine structures by the writer and N. S. Grace.[9]

[9] M. F. Crawford and N. S. Grace, Phys. Rev. **47**, 536 (1935).

This investigation was carried out under the supervision of Professor G. Breit, and I wish to thank him for the invaluable advice and assistance so freely given. I also take this opportunity to acknowledge the award of a Fellowship by the Royal Society of Canada, and to thank the University of Wisconsin and the Department of Physics for the privilege of working here.

MAY 15, 1935 PHYSICAL REVIEW VOLUME 47

Can Quantum-Mechanical Description of Physical Reality Be Considered Complete?

A. Einstein, B. Podolsky and N. Rosen, *Institute for Advanced Study, Princeton, New Jersey*

(Received March 25, 1935)

In a complete theory there is an element corresponding to each element of reality. A sufficient condition for the reality of a physical quantity is the possibility of predicting it with certainty, without disturbing the system. In quantum mechanics in the case of two physical quantities described by non-commuting operators, the knowledge of one precludes the knowledge of the other. Then either (1) the description of reality given by the wave function in quantum mechanics is not complete or (2) these two quantities cannot have simultaneous reality. Consideration of the problem of making predictions concerning a system on the basis of measurements made on another system that had previously interacted with it leads to the result that if (1) is false then (2) is also false. One is thus led to conclude that the description of reality as given by a wave function is not complete.

1.

Any serious consideration of a physical theory must take into account the distinction between the objective reality, which is independent of any theory, and the physical concepts with which the theory operates. These concepts are intended to correspond with the objective reality, and by means of these concepts we picture this reality to ourselves.

In attempting to judge the success of a physical theory, we may ask ourselves two questions: (1) "Is the theory correct?" and (2) "Is the description given by the theory complete?" It is only in the case in which positive answers may be given to both of these questions, that the concepts of the theory may be said to be satisfactory. The correctness of the theory is judged by the degree of agreement between the conclusions of the theory and human experience. This experience, which alone enables us to make inferences about reality, in physics takes the form of experiment and measurement. It is the second question that we wish to consider here, as applied to quantum mechanics.

Whatever the meaning assigned to the term *complete*, the following requirement for a complete theory seems to be a necessary one: *every element of the physical reality must have a counterpart in the physical theory*. We shall call this the condition of completeness. The second question is thus easily answered, as soon as we are able to decide what are the elements of the physical reality.

The elements of the physical reality cannot be determined by *a priori* philosophical considerations, but must be found by an appeal to results of experiments and measurements. A comprehensive definition of reality is, however, unnecessary for our purpose. We shall be satisfied with the following criterion, which we regard as reasonable. *If, without in any way disturbing a system, we can predict with certainty* (*i.e.*, *with probability equal to unity*) *the value of a physical quantity, then there exists an element of physical reality corresponding to this physical quantity.* It seems to us that this criterion, while far from exhausting all possible ways of recognizing a physical reality, at least provides us with one such way, whenever the conditions set down in it occur. Regarded not as a necessary, but merely as a sufficient, condition of reality, this criterion is in agreement with classical as well as quantum-mechanical ideas of reality.

To illustrate the ideas involved let us consider the quantum-mechanical description of the behavior of a particle having a single degree of freedom. The fundamental concept of the theory is the concept of *state*, which is supposed to be completely characterized by the wave function ψ, which is a function of the variables chosen to describe the particle's behavior. Corresponding to each physically observable quantity A there is an operator, which may be designated by the same letter.

If ψ is an eigenfunction of the operator A, that is, if

$$\psi' \equiv A\psi = a\psi, \tag{1}$$

where a is a number, then the physical quantity A has with certainty the value a whenever the particle is in the state given by ψ. In accordance with our criterion of reality, for a particle in the state given by ψ for which Eq. (1) holds, there is an element of physical reality corresponding to the physical quantity A. Let, for example,

$$\psi = e^{(2\pi i/h)p_0 x}, \tag{2}$$

where h is Planck's constant, p_0 is some constant number, and x the independent variable. Since the operator corresponding to the momentum of the particle is

$$p = (h/2\pi i)\partial/\partial x, \tag{3}$$

we obtain

$$\psi' = p\psi = (h/2\pi i)\partial\psi/\partial x = p_0\psi. \tag{4}$$

Thus, in the state given by Eq. (2), the momentum has certainly the value p_0. It thus has meaning to say that the momentum of the particle in the state given by Eq. (2) is real.

On the other hand if Eq. (1) does not hold, we can no longer speak of the physical quantity A having a particular value. This is the case, for example, with the coordinate of the particle. The operator corresponding to it, say q, is the operator of multiplication by the independent variable. Thus,

$$q\psi = x\psi \neq a\psi. \tag{5}$$

In accordance with quantum mechanics we can only say that the relative probability that a measurement of the coordinate will give a result lying between a and b is

$$P(a,b) = \int_a^b \bar{\psi}\psi dx = \int_a^b dx = b - a. \tag{6}$$

Since this probability is independent of a, but depends only upon the difference $b-a$, we see that all values of the coordinate are equally probable.

A definite value of the coordinate, for a particle in the state given by Eq. (2), is thus not predictable, but may be obtained only by a direct measurement. Such a measurement however disturbs the particle and thus alters its state. After the coordinate is determined, the particle will no longer be in the state given by Eq. (2). The usual conclusion from this in quantum mechanics is that *when the momentum of a particle is known, its coordinate has no physical reality.*

More generally, it is shown in quantum mechanics that, if the operators corresponding to two physical quantities, say A and B, do not commute, that is, if $AB \neq BA$, then the precise knowledge of one of them precludes such a knowledge of the other. Furthermore, any attempt to determine the latter experimentally will alter the state of the system in such a way as to destroy the knowledge of the first.

From this follows that either (1) *the quantum-mechanical description of reality given by the wave function is not complete* or (2) *when the operators corresponding to two physical quantities do not commute the two quantities cannot have simultaneous reality.* For if both of them had simultaneous reality—and thus definite values—these values would enter into the complete description, according to the condition of completeness. If then the wave function provided such a complete description of reality, it would contain these values; these would then be predictable. This not being the case, we are left with the alternatives stated.

In quantum mechanics it is usually assumed that the wave function *does* contain a complete description of the physical reality of the system in the state to which it corresponds. At first sight this assumption is entirely reasonable, for the information obtainable from a wave function seems to correspond exactly to what can be measured without altering the state of the system. We shall show, however, that this assumption, together with the criterion of reality given above, leads to a contradiction.

2.

For this purpose let us suppose that we have two systems, I and II, which we permit to interact from the time $t=0$ to $t=T$, after which time we suppose that there is no longer any interaction between the two parts. We suppose further that the states of the two systems before $t=0$ were known. We can then calculate with the help of Schrödinger's equation the state of the combined system I+II at any subsequent time; in particular, for any $t>T$. Let us designate the corresponding wave function by Ψ. We cannot, however, calculate the state in which either one of the two systems is left after the interaction. This, according to quantum mechanics, can be done only with the help of further measurements, by a process known as the *reduction of the wave packet*. Let us consider the essentials of this process.

Let $a_1, a_2, a_3, \cdots$ be the eigenvalues of some physical quantity A pertaining to system I and $u_1(x_1), u_2(x_1), u_3(x_1), \cdots$ the corresponding eigenfunctions, where x_1 stands for the variables used to describe the first system. Then Ψ, considered as a function of x_1, can be expressed as

$$\Psi(x_1, x_2) = \sum_{n=1}^{\infty} \psi_n(x_2) u_n(x_1), \tag{7}$$

where x_2 stands for the variables used to describe the second system. Here $\psi_n(x_2)$ are to be regarded merely as the coefficients of the expansion of Ψ into a series of orthogonal functions $u_n(x_1)$. Suppose now that the quantity A is measured and it is found that it has the value a_k. It is then concluded that after the measurement the first system is left in the state given by the wave function $u_k(x_1)$, and that the second system is left in the state given by the wave function $\psi_k(x_2)$. This is the process of reduction of the wave packet; the wave packet given by the infinite series (7) is reduced to a single term $\psi_k(x_2)u_k(x_1)$.

The set of functions $u_n(x_1)$ is determined by the choice of the physical quantity A. If, instead of this, we had chosen another quantity, say B, having the eigenvalues $b_1, b_2, b_3, \cdots$ and eigenfunctions $v_1(x_1), v_2(x_1), v_3(x_1), \cdots$ we should have obtained, instead of Eq. (7), the expansion

$$\Psi(x_1, x_2) = \sum_{s=1}^{\infty} \varphi_s(x_2) v_s(x_1), \tag{8}$$

where φ_s's are the new coefficients. If now the quantity B is measured and is found to have the value b_r, we conclude that after the measurement the first system is left in the state given by $v_r(x_1)$ and the second system is left in the state given by $\varphi_r(x_2)$.

We see therefore that, as a consequence of two different measurements performed upon the first system, the second system may be left in states with two different wave functions. On the other hand, since at the time of measurement the two systems no longer interact, no real change can take place in the second system in consequence of anything that may be done to the first system. This is, of course, merely a statement of what is meant by the absence of an interaction between the two systems. Thus, *it is possible to assign two different wave functions* (in our example ψ_k and φ_r) *to the same reality* (the second system after the interaction with the first).

Now, it may happen that the two wave functions, ψ_k and φ_r, are eigenfunctions of two noncommuting operators corresponding to some physical quantities P and Q, respectively. That this may actually be the case can best be shown by an example. Let us suppose that the two systems are two particles, and that

$$\Psi(x_1, x_2) = \int_{-\infty}^{\infty} e^{(2\pi i/h)(x_1 - x_2 + x_0)p} dp, \tag{9}$$

where x_0 is some constant. Let A be the momentum of the first particle; then, as we have seen in Eq. (4), its eigenfunctions will be

$$u_p(x_1) = e^{(2\pi i/h)px_1} \tag{10}$$

corresponding to the eigenvalue p. Since we have here the case of a continuous spectrum, Eq. (7) will now be written

$$\Psi(x_1, x_2) = \int_{-\infty}^{\infty} \psi_p(x_2) u_p(x_1) dp, \quad (11)$$

where

$$\psi_p(x_2) = e^{-(2\pi i/h)(x_2 - x_0)p}. \quad (12)$$

This ψ_p however is the eigenfunction of the operator

$$P = (h/2\pi i)\partial/\partial x_2, \quad (13)$$

corresponding to the eigenvalue $-p$ of the momentum of the second particle. On the other hand, if B is the coordinate of the first particle, it has for eigenfunctions

$$v_x(x_1) = \delta(x_1 - x), \quad (14)$$

corresponding to the eigenvalue x, where $\delta(x_1 - x)$ is the well-known Dirac delta-function. Eq. (8) in this case becomes

$$\Psi(x_1, x_2) = \int_{-\infty}^{\infty} \varphi_x(x_2) v_x(x_1) dx, \quad (15)$$

where

$$\varphi_x(x_2) = \int_{-\infty}^{\infty} e^{(2\pi i/h)(x - x_2 + x_0)p} dp$$
$$= h\delta(x - x_2 + x_0). \quad (16)$$

This φ_x, however, is the eigenfunction of the operator

$$Q = x_2 \quad (17)$$

corresponding to the eigenvalue $x + x_0$ of the coordinate of the second particle. Since

$$PQ - QP = h/2\pi i, \quad (18)$$

we have shown that it is in general possible for ψ_k and φ_r to be eigenfunctions of two noncommuting operators, corresponding to physical quantities.

Returning now to the general case contemplated in Eqs. (7) and (8), we assume that ψ_k and φ_r are indeed eigenfunctions of some noncommuting operators P and Q, corresponding to the eigenvalues p_k and q_r, respectively. Thus, by measuring either A or B we are in a position to predict with certainty, and without in any way disturbing the second system, either the value of the quantity P (that is p_k) or the value of the quantity Q (that is q_r). In accordance with our criterion of reality, in the first case we must consider the quantity P as being an element of reality, in the second case the quantity Q is an element of reality. But, as we have seen, both wave functions ψ_k and φ_r belong to the same reality.

Previously we proved that either (1) the quantum-mechanical description of reality given by the wave function is not complete or (2) when the operators corresponding to two physical quantities do not commute the two quantities cannot have simultaneous reality. Starting then with the assumption that the wave function does give a complete description of the physical reality, we arrived at the conclusion that two physical quantities, with noncommuting operators, can have simultaneous reality. Thus the negation of (1) leads to the negation of the only other alternative (2). We are thus forced to conclude that the quantum-mechanical description of physical reality given by wave functions is not complete.

One could object to this conclusion on the grounds that our criterion of reality is not sufficiently restrictive. Indeed, one would not arrive at our conclusion if one insisted that two or more physical quantities can be regarded as simultaneous elements of reality *only when they can be simultaneously measured or predicted.* On this point of view, since either one or the other, but not both simultaneously, of the quantities P and Q can be predicted, they are not simultaneously real. This makes the reality of P and Q depend upon the process of measurement carried out on the first system, which does not disturb the second system in any way. No reasonable definition of reality could be expected to permit this.

While we have thus shown that the wave function does not provide a complete description of the physical reality, we left open the question of whether or not such a description exists. We believe, however, that such a theory is possible.

A.2 Appendix—Arguing with the Scholastic Method

Using Ockham's discussion on universals from his Opera Theologica II (of Ockham 2012, OT II, Liber I, D.2 Q.4, p. 100) the structure of the scholastic tradition can be seen quite well in the following example.

We follow the translation from Spade (1994). Ockham there poses the question:

Question 4: Is a universal a thing outside the soul, in individuals, really distinct from them, and not multiplied in individuals?

Then he argues in favour of the position that affirms the question:

(1) As for the identity and distinction of God from creature, it must be asked whether there is something univocal common to God and creature and essentially predicable of both. But because this question, along with much of what has already been said and is about to be said in the following questions, depends on a knowledge of univocal and universal nature, there fore in order to clarify what has been and will be said I will first ask some questions about universal and univocal nature.
(2) On this, I first ask whether what is immediately and proximately denominated by a universal and univocal intention [K Ch. 23] is truly some thing outside the soul, intrinsic and essential to what it is common and univocal to, and really distinct from them. (3) Yes it is:
(4) First, it is truly a thing, essential and intrinsic to what it is common to.
...

He then presents the view of his opposer, who he calls *"Doctori Subtili"* (of Ockham 2012, OT II, Liber I, D.2 Q.4, p. 100). Some researchers assume that he refers to Duns Scotus, and some would argue that the view he presents would better represent Walter Burleys ideas (Spade 1994, p. 115):

Walter Burleys view - Opinio Doctori Subtili Imposita

(8) On this question there is one theory that says every univocal universal is a certain thing existing outside the soul, really in each singular and belong ing to the essence of each singular, really distinct from each singular and from any other universal, in such a way that the universal man is truly one thing outside the soul, existing really in each man, and is really distinguished from each man and from the universal animal and from the universal sub stance.
...

Following an extensive argumentation in favour on this view a not less extensive list of arguments against it is presented:

Against this theory - Contra Opinonem Doctori Subtili Impositam

(34) This view is absolutely false and absurd. So I argue against it.
(35) First: No thing one in number, not varied or multiplied, is in several sensible supposita or singulars, or for that matter in any created individuals, at one and the same time. But a thing such as this theory postulates, if it were granted, would be one in number. Therefore, it would not be in several singulars and belong to their essence.
...

Now Ockham presents his own views

Ockham's own theory - Opinio Auctoris

(95) Therefore, in reply to the question [(2)], I say otherwise: No thing really distinct from singular things and intrinsic to them is universal or common to them. For such a thing is not to be posited except (a) to preserve the one's essential predication of the other, or (b) to preserve our knowledge of things and (c) the definitions of things. Aristotle, [Metaphysics XIII.4, 1078,27-34], suggests these reasons for Plato's theory.
...

followed by replies to the before presented theories:

Replies to the arguments in favor of Burley's theory - Ad Argumenta Opinionis Doctori Subtili Impositae

(111) To the other theory's first argument

...

and

Reply to the preliminary arguments - Responsio Ad Argumenta Principalia

(184) To the first main argument [(4)], I reply: The universal man and a particular man are essentially one is literally false.

...

The scholastic method is a completely different way of arguing than what we are used to nowadays. To modern eyes, it appears in some sense excessive and inefficient, though it prepared the ground for modern scientific reasoning, which is built on hypotheses, induction, deduction, and discussion.

For more details concerning the scholastic method, refer to Kretzmann et al. (1982). There, a summary and detailed presentation of the scholastic method and its historical embedding is given.

A.3 Appendix—Modern Views on the Contradiction Principle, the Notion of Identity and Indistinguishability

Ockham's Contradiction principle has some predecessors and modern representations, and pervaded the history of philosophy in different representations and names. It would exceed the scope of this paper to give an overview that could claim to be complete. Therefore, a certain selection is presented that seems particular relevant to the author in context of this discussion to demonstrate the application of Ockham's razor.

Whitehead and Russell are prominent for their attempt to put mathematics on a logical basis. In this canon, the law of Non-Contradiction also found its place Whitehead and Russell (1912, p. 117 *3,24):

$$\vdash \neg(p \wedge \neg p)$$

Two contradictory propositions cannot both be true. In the context of propositions about the properties of objects, difference can be posed through contradiction. This argument and application of Contradiction is found later in Wittgenstein (2003) which refers to this principle several times. See Wittgenstein (2003, 2.0233 and 2.02331).

If two objects have the same logical form, the only distinction between them, apart from their external properties, is that they are different.

> *Either a thing has properties that nothing else has, in which case we can immediately use a description to distinguish it from the others and refer to it; or, on the other hand, there are several things that have the whole set of their properties in common, in which case it is quite impossible to indicate one of them. For if there is nothing to distinguish a thing, I cannot distinguish it, since otherwise it would be distinguished after all.*

or at a different place (Wittgenstein 2003, 5.53, 5.5301, 5.5302, 5.5303):

> *Identity of the object I express by identity of the sign and not by means of a sign of identity. Difference of the objects by difference of the signs.*

> *It is self-evident that identity is not a relation between objects. This becomes very clear if one considers, for example, the proposition* $(x) : fx. \subset .x = a$. *What this proposition says is simply that only a satisfies the function f, and not that only things that have a certain relation to a satisfy the function. Of course, it might then be said that only a did have this relation to a; but in order to express that, we should need the identity-sign itself.*

> *Russell's definition of '=' is inadequate, because according to it we cannot say that two objects have all their properties in common. (Even if this proposition is never correct, it still has sense.)*

> *Roughly speaking, to say of two things that they are identical is nonsense, and to say of one thing that it is identical with itself is to say nothing at all.*

The laws of contradiction are always in connection with the question of identity of objects, beginning with the pre-Socratics and their philosophy of substance and matter. Aristotle based this and, of course, his own philosophical ideas on logical foundations that still pervade our thinking. In classical physics, the identity of physical objects seemed clearly defined, but the foundational questions have there have been already covered by the operational success Newtonian Mechanics offered. When quantum mechanics arose, those philosophical questions became significant again in physics because of their sudden experimental accessibility and support by measurements on the microscopical scale.

The individuality and nature of quantum objects became a striking question of quantum philosophy, starting at the statistical mechanics involving quantum mechanics. By neglecting the identity of permuted particles, the experimentally confirmed laws of statistical mechanics have been found. It is an interesting observation that physics and philosophy develop in a kind of entangled way and stimulate each other.

Quantum mechanics kept the topic of identity challenging up to now, as can be seen by ongoing discussions on the definition and interpretation of identity, individuality and indistinguishability (Krause and Arenhart 2014; Da Costa and Lombardi 2014; Dieks 2014). For an account of individuality in the Bohm interpretation, refer to Pylkkänen et al. (2014).

A.4 Appendix—Interpretations of Quantum Mechanics

We want to give a brief overview of the variety of interpretations of quantum mechanics and the involved ontology.

Copenhagen Interpretation

The Copenhagen interpretation was the first interpretation of quantum mechanics, but there exist several versions that differ in subtleties. In general quantum mechanics is seen as a complete description of nature. Measurements are described by the projection postulate. Heisenberg emphasized the discontinuous evolution from potentiality to actuality during the measurement process. The statistical character and its application to a single quantum system discerns different versions of the Copenhagen interpretation.

Ensemble Interpretation

The ensemble interpretation proposes that quantum mechanics refers to a statistical ensemble of systems and nothing can be said about the properties and behaviour of a single quantum system.

Informational Interpretation—Brukner and Zeilinger

The Brukner-Zeilinger interpretation is a continuation of the Copenhagen interpretation with an emphasis on the concept of information as a foundational resource and principle (Zeilinger 2002, p. 252):

> *... it is suggested that information is the most basic notion of quantum mechanics, and it is information about possible measurement results that is represented in the quantum states.*

On the ontological level, information is seen as the furniture of the universe. It is seen as primary, even to the spacetime framework, which is seen also as an aspect of information. The conception of information drastically differs from the notions of in-formation in the De Broglie-Bohm interpretation and information in QBism.

QBism

QBism is a continuation of the Copenhagen interpretation. It is the attempt to reconstruct quantum mechanics on the basis of probabilities as representation of the subjective degrees of belief of agents interacting with the world. The quantum state is not seen as a foundational entity; it is merely a different way to represent the probabilities. The Hilbert space is constructed out of SIC-POVMs, which are symmetric, informationally complete, positive operator-valued measures. They generate the probabilities for the individual agents based on their subjective beliefs according to the fundamental postulate that reproduces the Born rule for the probabilities.

QBism is a local theory, takes a realistic position and provides an epistemic interpretation of the wavefunction. In the context with structural realism, we proposed that QBism should be seen as doxastic structural realism, since its subjective perspective distinguished QBism's position from other interpretations.

Modal Interpretations

The modal interpretation includes a class of interpretations. The basic idea is that not all observables of a quantum system have definite values. Furthermore two states are applied to a quantum system; a dynamical state and a value state. It is a realistic interpretation of quantum mechanics in the sense of defined properties of the system (Lombardi and Dieks 2017):

> *The interpretation should be "realist" in the precise sense that it assumes that quantum systems always possess a number of definite properties, which may change with time. It should be noted, however, that this semantic realism is compatible with agnosticism or van Fraassen's brand of empiricism (van Fraassen 1991, Bueno 2014), and does not presuppose epistemological realism.*

Quantum Logic

Quantum logic modifies the rules of classical boolean logic to give consideration to the behaviour of quantum systems (Mittelstaedt 1978).

Relative State Interpretations and Many Worlds

The relative state interpretation, or Many worlds interpretation, avoids to use additional postulates like the projection postulate in Copenhagen interpretation, or the guidance equation in Bohmian mechanics. It interprets the wave-function as not being subject to any wave-function reduction. To every branch of a universal wave-function it assigns equally reality and states that the world is in a superposition of many parallel worlds. Since its original proposal by Everett III (1973) several versions and continuations developed.

Mathematical Universe Hypothesis

The Mathematical Universe Hypothesis is not a stand-alone interpretation of quantum mechanics, since it is related to the Many worlds interpretation. It proposes mathematical objects to be the foundational ontological entities.

Bohmian Mechanics

Bohmian mechanics in the reading of Dürr and Teufel (2009) is a theory that extends the formalism of quantum mechanics by an additional postulate and thereby provides a particle ontology. Bohmian mechanics is related to the De Broglie-Bohm Interpretation in the sense that the guidance equation is identical, but the De Broglie-Bohm Interpretation concludes the guidance equation out of interpretation of the Schröedinger equation, whereas Bohmian mechanics postulates the guidance equation and shows the agreement of the theory with empirical results. The ontology involves particles that have well-defined positions but do not carry properties in the sense of classical objects.

De Broglie-Bohm Interpretation

The De Broglie-Bohm Interpretation is based on a reinterpretation of the Schrödinger equation that results in two real-valued equations that resemble the Hamilton-Jacobi formalism of classical mechanics with an additional term. Out of this, the guidance equation is derived. The De Broglie-Bohm Interpretation is a nonlocal realistic hidden-variable interpretation, although it is far from classical physics (Holland 1995, p. 328):

> *The classical ideal of passive measurements which simply reveal a preexisting reality cannot be sustained when we come to the quantum treatment of this problem.*

From the perspective of structural realism, the De Broglie Bohm interpretation emphasizes the relation of quantum objects, encoded in the quantum potential and the wave function. If one asks for the interpretation of the wave function from this perspective, an appropriate answer would be that the wave function encodes the relations of the involved quantum objects.

Consistent Histories

The consistent histories approach is a continuation of the Copenhagen interpretation. It involves reasoning principles that differ from classical logic (Griffiths 1984). It emphasizes the statistical character of the formalism, abandons the concept of uniquely defined properties of quantum systems and proposes an ambiguous reality (Griffiths 2003, p. 317):

> *... reality is such that it can be described in various alternative, incompatible ways, using descriptions which cannot be combined or compared.*

Transactional Interpretation

The transactional interpretation of quantum mechanics emphasizes the wave function and its complex conjugate as real and interprets them as waves running forward and backward in time (Cramer 2016).

It resembles the two-state vector formalism, which also emphasizes the wave function and its conjugate as real.

Cellular Automaton Interpretation

The Cellular automaton Interpretation is the attempt to demonstrate that quantum mechanics can be reformulated without Hilbert space formalism. It is a reconstruction of quantum mechanics in terms of a cellular automata (Hooft 2014).

Relational Interpretation

Rovelli proposed an interpretation of quantum mechanics names Relational quantum mechanics(RQM) based on the notion of events and their relative relation (Rovelli 2016, p. 1235):

> *What does produce the events we observe, in RQM? The answer is "nothing": they are the primitive ontology of the theory. Asking what produces them is like asking what produces the particles or the fields of the Standard Model. In the theory, these are not tot to be explained: they are the ingredients to explain the rest.*

Rovelli emphasizes the discreteness of the quantum events as fundamental. The quantum event is the foundational ontology of this interpretation.

No Interpretation by Fuchs and Peres

Fuchs and Peres advance the position that it is superfluous to interpret quantum mechanics beyond its empirical prediction, in which quantum mechanics is more than successful and accurate (Fuchs and Peres 2000). They do not see the merit of the manifold of interpretations and their ontological commitments (Fuchs and Peres 2000, p. 9):

> *We do not find any merit in the various alternatives that were proposed to the straightforward interpretation of quantum theory: It is a set of rules for calculating probabilities for macroscopic detection events, upon taking into account any previous experimental information.*

Bohr-Ludwigs Reconstruction of Quantum Mechanics

Günther Ludwig provided a reformulation of quantum mechanics. Ludwig's approach is built upon sets that allow to define unsharp measurements without the operator formalism and referred to the statistical character of quantum mechanics in the sense of ensembles (Ludwig and Thurler 2007).

Växjo Interpretation

The Växjo interpretation takes a position combining the positions of Einstein and Bohr by admitting at the same time that quantum mechanics is not complete, since it is a statistical description, and that quantum mechanics is fundamentally contextual (Khrennikov et al. 2012). It proposes that quantum mechanics describes correlations in prequantum random fields.

CBH Theorem

The CBH theorem is a reconstruction of quantum mechanics based on informational principles (Clifton et al. 2003). The proposed principles differ from the principles advanced in the Brukner-Zeilinger interpretation.

Collapse Models

The family of collapse models includes several models that claim to solve the measurement problem in quantum mechanics by a process to explain the projection postulate. They are not interpretations of quantum mechanics, since they alter the formalism of quantum mechanics by adding stochastic or nonlinear mechanisms. Since they claim to solve the measurement problem and propose an ontology, they are closely related to the interpretations of quantum mechanics.

A Map of Madness

Cabello (2015) presented a map for interpretations of quantum mechanics that included most interpretations of quantum mechanics. We want to amend this map by the De Broglie-Bohm interpretation and propose to use the term doxastic for statements on reconstructions or interpretations of quantum mechanics that refer to the concept of belief (Table A.1).

Table A.1 A map of madness

Type I		Type II	
Ψ-ontic	Ψ-epistemic	Epistemic	Doxastic
Bohmian mechanics De Broglie-Bohm Many worlds Modal Bell's beables Collapse theories	Einstein Ballentine Consistent histories Spekkens	Copenhagen Wheeler Relational Zeilinger-Brukner No interpretation	QBism

Table A.2 An extended map of madness

	Type I		Type II	
	Ψ-ontic	Ψ-epistemic	Epistemic	Doxastic
Reductionism	Modal Bell's beables Collapse theories	Einstein Ballentine? Consist. hist. Spekkens?	Copenhagen Wheeler Relational Zeilinger? Brukner? No interpr.	QBism?
Holism	Bohm. mechanics De Broglie-Bohm Many worlds			

Furthermore we want to propose to add further ontological characteristics to the structure of the map to distinguish interpretations concerning these characteristics. The analysis for several of these characteristics for all interpretations is subject to future work.

There are several dichotomies that are of interest in that context and the analysis for all interpretations is subject to future work: "locality–nonlocality", "determinism–indeterminism", "nominalism–platonism", "objectivity–subjectivity", "realism–antirealism", "object emphasis–structure emphasis."

In the first place this has been done for the dichotomy of "reductionism–holism." Since reductionism and holism are itself complex conceptions that can refer to the ontological level or the theory levels, we interpret them in a wider sense. If the theory includes a stance that tends in the one or the other direction it will be assigned with the specific conception. If the interpretation is ambiguous, neutral, or not yet analysed we assigned a question mark behind (Table A.2).

What is discussible is the notion of intrinsic properties. If one assumes the Hilbert space dimension of the quantum system to be an intrinsic property, then QBism for instance would not be a Type II interpretation. On the other hand it is hard to deny that the Hilbert space dimension is in some sense intrinsic to the quantum system. We propose that this asks for clarification of the used terms. We leave this question open for future works.

A.5 Appendix—Quambiguity Object

For the purpose of presentations and talks, we designed an object that demonstrates the conception of Quambiguity hands-on. The letter K stands for Kopenhagen-like interpretations, the letter B for Bohmian mechanics and De-Broglie Bohm interpretation, and the letter M for the Many-worlds interpretation.

References

Achinstein, P. (1991). *Particles and waves: Historical essays in the philosophy of science*. Oxford University Press.
Andreas, H. (2016). Theoretical terms in science. In Zalta, E. N., editor, *The Stanford Encyclopedia of Philosophy*. Metaphysics Research Lab, Stanford University, winter 2016 edition.
Andreas, H. and Zenker, F. (2014). Basic concepts of structuralism. *Erkenntnis*, 79(8):1367–1372.
Avenarius, R. (1906). Kritik der reinen erfahrung.
Avenarius, R. and Avenarius, R. H. L. (1905). *Der menschliche Weltbegriff*. OR Reisland.
Balzer, W. (1980). Günther ludwigs grundstrukturen einer physikalischen theorie. *Erkenntnis*, 15(3):391–408.
Beckmann, J. P. (1995). *Wilhelm von Ockham*, volume 533. CH Beck.
Beckmann, J. P. (1998). Ockham, ockhamismus, und nominalismus: Spuren der wirkungsgeschichte des venerabilis inceptors. *Franciscan Studies*, 56:77–95.
Bell, J. S. (1964). On the Einstein-Podolsky-Rosen paradox. *Physics*, 1:195–200.
Bell, J. S. (1981). Bertlmann's socks and the nature of reality. *Le Journal de Physique Colloques*, 42(C2):C2–41.
Bohm, D. and Curd, M. (1981). Wholeness and the implicate order.
Bohm, D. and Hiley, B. J. (1984). Measurement understood through the quantum potential approach. *Foundations of Physics*, 14(3):255–274.
Bohm, D., Hiley, B. J., and Kaloyerou, P. N. (1987). An ontological basis for the quantum theory. *Physics Reports*, 144(6):321–375.
Bohr, N. (1935). Can quantum-mechanical description of physical reality be considered complete? *Physical review*, 48(8):696.
Braver, L. (2007). *A thing of this world: A history of continental anti-realism*. Northwestern University Press.
Brukner, Č. (2017). On the quantum measurement problem. In *Quantum [Un] Speakables II*, pages 95–117. Springer.
Brukner, Č. and Zeilinger, A. (1999). Operationally invariant information in quantum measurements. *Physical Review Letters*, 83(17):3354.
Brukner, Č. and Zeilinger, A. (2003). Information and fundamental elements of the structure of quantum theory. In *Time, quantum and information*, pages 323–354. Springer.
Cabello, A. (2015). Interpretations of quantum theory: a map of madness. *arXiv preprint* arXiv:1509.04711.
Callender, C. (2015). One world, one beable. *Synthese*, 192(10):3153–3177.

G. C. Krizek, *The Conceptions of Reality in the Interpretations of Quantum Mechanics*,
https://doi.org/10.1007/978-3-031-65786-3

Carlson, M. (2015). Logic and the structure of the web of belief. *Journal for the History of Analytical Philosophy*, 3(5).
Carnap, R. (1923). Über die aufgabe der physik. *Kant-Studien*, 28(1-2):90–107.
Chalmers, A. (2009). *The scientist's atom and the philosopher's stone: How science succeeded and philosophy failed to gain knowledge of atoms*, volume 279. Springer.
Chalmers, A. (2014). Atomism from the 17th to the 20th century. In Zalta, E. N., editor, *The Stanford Encyclopedia of Philosophy*. Metaphysics Research Lab, Stanford University, winter 2014 edition.
Clauser, J. F., Horne, M. A., Shimony, A., and Holt, R. A. (1969). Proposed experiment to test local hidden-variable theories. *Phys. Rev. Lett.*, 23:880–884.
Clifton, R., Bub, J., and Halvorson, H. (2003). Characterizing quantum theory in terms of information-theoretic constraints. *Foundations of Physics*, 33(11):1561–1591.
Combourieu, M.-C. and Rauch, H. (1992). The wave-particle dualism in 1992: A summary. *Foundations of physics*, 22(12):1403–1434.
Cramer, J. G. (2016). *The Quantum Handshake: Entanglement, Nonlocality and Transactions*. Springer International Publishing.
Cushing, J. T., Fine, A., and Goldstein, S. (2013). *Bohmian mechanics and quantum theory: an appraisal*, volume 184. Springer Science & Business Media.
Da Costa, N. and Lombardi, O. (2014). Quantum mechanics: Ontology without individuals. *Foundations of Physics*, 44(12):1246–1257.
Da Costa, N. C. and French, S. (1990). The model-theoretic approach in the philosophy of science. *Philosophy of Science*, 57(2):248–265.
Dawid, R. (2007). Scientific realism in the age of string theory. *Physics and Philosophy*.
Dawid, R. (2008). Scientific prediction and the underdetermination of scientific theory building.
Dawid, R. (2015). Modelling scientific confirmation.
Dawid, R., Hartmann, S., and Sprenger, J. (2014). The no alternatives argument. *The British Journal for the Philosophy of Science*, page axt045.
Dennis, G., de Gosson, M. A., and Hiley, B. J. (2015). Bohm's quantum potential as an internal energy. *Physics Letters A*, 379(18):1224–1227.
Di Domenico, A., Gabriel, A., Hiesmayr, B. C., Hipp, F., Huber, M., Krizek, G., Mühlbacher, K., Radic, S., Spengler, C., and Theussl, L. (2012). Heisenberg's uncertainty relation and bell inequalities in high energy physics. *Foundations of Physics*, 42(6):778–802.
Dieks, D. (2014). The logic of identity: Distinguishability and indistinguishability in classical and quantum physics. *Foundations of Physics*, 44(12):1302–1316.
Dürr, D. and Teufel, S. (2009). *Bohmian mechanics: the physics and mathematics of quantum theory*. Springer Science & Business Media.
Einstein, A. (1933). Zur methode der theoretischen physik - herbert spencer lecture, oxford. Albert Einstein Archives, The Hebrew University of Jerusalem, Israel. 1-114.
Einstein, A. (1934). On the method of theoretical physics. *Philosophy of science*, 1(2):163–169.
Einstein, A. (1936). Physics and reality. *Journal of the Franklin Institute*, 221(3):313–347.
Einstein, A. (1946). *Remarks on Bertrand Russell's theory of knowledge*. Library of Living Philos.
Einstein, A. (1948). Quantum mechanics and reality. *Dialectica*, 2(3-4):320–324.
Einstein, A., Infeld, L., Preusser, W., and Fölsing, A. (1950). *Die Evolution der Physik*. Zsolnay Hamburg.
Einstein, A., Podolsky, B., and Rosen, N. (1935). Can quantum-mechanical description of physical reality be considered complete? *Physical review*, 47(10):777.
Einstein, A. and Rosen, N. (1935a). Das partikel-problem von standpunkte der allgemeinen relativitätstheorie. Einstein archives online. Archival Call Number: 1-130.
Einstein, A. and Rosen, N. (1935b). The particle problem in the general theory of relativity. *Physical Review*, 48(1):73.
Ellis, B. (2005). Physical realism. *Ratio*, 18(4):371–384.
Englert, B.-G. (2013). On quantum theory. *arXiv preprint* arXiv:1308.5290.

Esfeld, M. (2004). Quantum entanglement and a metaphysics of relations. *Studies in History and Philosophy of Science Part B: Studies in History and Philosophy of Modern Physics*, 35(4):601–617.

Esfeld, M. (2007). Metaphysics of science between metaphysics and science. *Grazer Philosophische Studien*, 74(1):199–213.

Esfeld, M. (2008). *Naturphilosophie als Metaphysik der Natur*. Suhrkamp Frankfurt am Main.

Esfeld, M. (2012). *Philosophie der Physik*. Suhrkamp.

Esfeld, M. (2013). Ontic structural realism and the interpretation of quantum mechanics. *European Journal for Philosophy of Science*, 3(1):19–32.

Esfeld, M., Hubert, M., Lazarovici, D., and Dürr, D. (2013). The ontology of bohmian mechanics. *The British Journal for the Philosophy of Science*, page axt019.

Esfeld, M. and Lam, V. (2008). Moderate structural realism about space-time. *Synthese*, 160(1):27–46.

Everett III, H. (1973). The theory of the universal wave function. In *The many-worlds interpretation of quantum mechanics*. Citeseer.

Fahrbach, L. (2009). *Pessimistic Meta-Induction and the Exponential Growth of Science*. na.

Feyerabend, P. (1993). *Against method*. Verso.

Fine, A. (2009). *The shaky game*. University of Chicago Press.

Fiske, S. T. and Taylor, S. E. (2013). *Social cognition: From brains to culture*. Sage.

Fuchs, C. A. (2010). Qbism, the perimeter of quantum bayesianism. *arXiv preprint* arXiv:1003.5209.

Fuchs, C. A. (2017a). Notwithstanding bohr, the reasons for qbism. *arXiv preprint* arXiv:1705.03483.

Fuchs, C. A. (2017b). On participatory realism. In *Information and Interaction*, pages 113–134. Springer.

Fuchs, C. A. and Peres, A. (2000). Quantum theory needs no 'interpretation'. *Physics Today*, 53(3):70–71.

Fuchs, C. A. and Stacey, B. C. (2016). Qbist quantum mechanics: Quantum theory as a hero's handbook. *arXiv preprint* arXiv:1612.07308.

Gähde, U. (2014). Theory-dependent determination of base sets: Implications for the structuralist approach. *Erkenntnis*, 79(8):1459–1473.

Gallie, W. B. (1955). Essentially contested concepts. In *Proceedings of the Aristotelian society*, volume 56, pages 167–198. JSTOR.

Gardner, M. R. (1972). Quantum-theoretical realism: Popper and einstein v. kochen and specker. *The British Journal for the Philosophy of Science*, 23(1):13–23.

Ghose, P. and Home, D. (1996). The two-prism experiment and wave-particle duality of light. *Foundations of Physics*, 26(7):943–953.

Gisin, N. (2013). Are there quantum effects coming from outside space–time? nonlocality, free will and "no many-worlds". In *Is Science Compatible with Free Will?*, pages 23–39. Springer.

Graney, C. M. (2010). The telescope against copernicus: Star observations by riccioli supporting a geocentric universe. *Journal for the History of Astronomy*, 41(4):453–467.

Greaves, H. (2007). Probability in the everett interpretation. *Philosophy Compass*, 2(1):109–128.

Griffiths, R. B. (1984). Consistent histories and the interpretation of quantum mechanics. *Journal of Statistical Physics*, 36(1-2):219–272.

Griffiths, R. B. (2003). *Consistent quantum theory*. Cambridge University Press.

Hardy, L. (2013). *Contextuality in Bohmian mechanics*. Springer Science & Business Media.

Harrigan, N. and Spekkens, R. W. (2010). Einstein, incompleteness, and the epistemic view of quantum states. *Foundations of Physics*, 40(2):125–157.

Heider, F. et al. (1958). The psychology of interpersonal relations. *The Journal of Marketing*, 56:322.

Held, C. (2012). Incompatibility of standard completeness and quantum mechanics. *International Journal of Theoretical Physics*, 51(9):2974–2984.

Held, C. (2015). Einstein's boxes: Incompleteness of quantum mechanics without a separation principle. *Foundations of Physics*, 45(9):1002–1018.

Hempel, C. (1977). Formulation and formalization of scientific theories. *The structure of scientific theories*, 2:244–55.
Hempel, C. G. (1970). On the" standard conception" of scientific theories. *Analyses of theories and methods of physics and psychology*.
Hentschel, K. (1986). Die korrespondenz einstein-schlick: Zum verhältnis der physik zur philosophie. *Annals of Science*, 43(5):475–488.
Hiesmayr, B. C. (2007). Nonlocality and entanglement in a strange system. *The European Physical Journal C*, 50(1):73–79.
Hilary, P. (1988). *Representation and reality*. MIT press.
Hiley, B. J. and Pylkkanen, P. (2005). Can mind affect matter via active information? *Mind and Matter*, 3(2):8–27.
Holland, P. R. (1995). *The quantum theory of motion: an account of the de Broglie-Bohm causal interpretation of quantum mechanics*. Cambridge university press.
Hooft, G. (2014). The cellular automaton interpretation of quantum mechanics. *arXiv preprint* arXiv:1405.1548.
Howard, D. (1993). Was einstein really a realist? *Perspectives on Science*, 1(2):204–251.
Howard, D. A. (2005). Albert einstein as a philosopher of science. *Physics today*, 58(12):34–40.
Ishigaki, T. (1995). A formal system for classical particle mechanics, its model-theoretic applications and space-time structure. *Synthese*, 102(2):267–292.
Jannes, G. (2009). Some comments on "the mathematical universe". *Foundations of Physics*, 39(4):397–406.
Janssen, M. and Lehner, C. (2014). *The Cambridge Companion to Einstein*, volume 1. Cambridge University Press.
Kallfelz, F. (1929). *Das Ökonomieprinzip bei Ernst Mach*. J.B.Lindl.
Kant, I. (1889). *Immanuel Kants Kritik der reinen Vernunft*. Mayer & Müller.
Kelly, K. T. (2011). Simplicity, truth, and probability. In *Philosophy of statistics*, pages 983–1024. Elsevier.
Khlentzos, D. (2016). Challenges to metaphysical realism. In Zalta, E. N., editor, *The Stanford Encyclopedia of Philosophy*. Metaphysics Research Lab, Stanford University, winter 2016 edition.
Khrennikov, A. (2016). Reflections on zeilinger–brukner information interpretation of quantum mechanics. *Foundations of Physics*, 46(7):836–844.
Khrennikov, A., Khrennikov, A., Atmanspacher, H., Migdall, A., and Polyakov, S. (2012). Växjö interpretation of wave function: 2012. In *AIP Conference Proceedings*, volume 1508, pages 244–252. AIP.
Kiefer, C. (2015). *Albert Einstein, Boris Podolsky, Nathan Rosen: Kann die quantenmechanische Beschreibung der physikalischen Realität als vollständig betrachtet werden?* Springer-Verlag.
Klocker, H. R. (1985). Ockham and the divine freedom. *Franciscan studies*, 45:245–261.
Kochen, S. and Specker, E. P. (1975). The problem of hidden variables in quantum mechanics. In *The Logico-Algebraic Approach to Quantum Mechanics*, pages 293–328. Springer.
Kofler, J. and Zeilinger, A. (2013). Quantum information and randomness. *arXiv preprint* arXiv:1301.2515.
Kox, A. J. (2014). Einstein on statistical physics - fluctuations and atomism. In *The Cambridge companion to Einstein*, pages 103–116. Cambridge University Press.
Krause, D. and Arenhart, J. R. (2014). Separability and non-individuality: Is it possible to conciliate (at least a form of) einstein's realism with quantum mechanics? *Foundations of Physics*, 44(12):1269–1288.
Kretzmann, N., Kenny, A., Pinborg, J., and Stump, E. (1982). *The Cambridge history of later medieval philosophy: from the rediscovery of Aristotle to the disintegration of scholasticism, 1100-1600*. Cambridge Univ Press.
Krizek, G. C. and Mairhofer, L. (2023). Notions of completeness in the epr discussion. *Entropy*, 25(4):585.

Kuhlmann, M. (2010). *The Ultimate Constituents of the material world: in search of an ontology for fundamental physics*, volume 37. Walter de Gruyter.
Kuhn, T. S. (1957). *The Copernican revolution: Planetary astronomy in the development of Western thought*, volume 16. Harvard University Press.
Kuhn, T. S. (1962). *The Structure of Scientific Revolutions*. University of Chicago Press.
Kupczynski, M. (2006). Seventy years of the epr paradox. *AIP Conference Proceedings*, 861(516).
Ladyman, J. (1998a). What is structural realism? *Studies in History and Philosophy of Science Part A*, 29(3):409 – 424.
Ladyman, J. (1998b). What is structural realism? *Studies in History and Philosophy of Science Part A*, 29(3):409–424.
Ladyman, J. (2016). Structural realism. In Zalta, E. N., editor, *The Stanford Encyclopedia of Philosophy*. Metaphysics Research Lab, Stanford University, winter 2016 edition.
Landsman, N. P. (2006). When champions meet: Rethinking the bohr–einstein debate. *Studies in History and Philosophy of Science Part B: Studies in History and Philosophy of Modern Physics*, 37(1):212–242.
Laudan, L. (1981). A confutation of convergent realism. *Philosophy of science*, 48(1):19–49.
Le Morvan, P. (2004). Arguments against direct realism and how to counter them. *American Philosophical Quarterly*, 41(3):221–234.
Lehner, C. (2014). Einstein's realism and his critique of quantum mechanics. In *The Cambridge companion to Einstein*, pages 306–353. Cambridge University Press.
Lombardi, O. and Dieks, D. (2017). Modal interpretations of quantum mechanics. In Zalta, E. N., editor, *The Stanford Encyclopedia of Philosophy*. Metaphysics Research Lab, Stanford University, spring 2017 edition.
Ludwig, G. (1978). *Die Grundstrukturen einer physikalischen Theorie*. Springer-Verlag.
Ludwig, G. and Thurler, G. (2007). *A new foundation of physical theories*. Springer Science & Business Media.
Mach, E. (1898). *Popular scientific lectures*. Open court publishing company.
Mach, E. (1922). *Analyse der Empfindungen*. Verlag von Georg Fischer.
Maldacena, J. and Susskind, L. (2013). Cool horizons for entangled black holes. *Fortschritte der Physik*, 61(9):781–811.
Mermin, N. D. (2012). Commentary: Quantum mechanics: Fixing the shifty split. *Physics Today*, 65(7):8–10.
Mittelstaedt, P. (1978). Quantum logic. In *Physical Theory as Logico-Operational Structure*, pages 153–166. Springer.
Morganti, M. (2004). On the preferability of epistemic structural realism. *Synthese*, 142(1):81–107.
Müller, F. (2015). The pessimistic meta-induction: Obsolete through scientific progress? *International Studies in the Philosophy of Science*, 29(4):393–412.
of Ockham, W. (2012). *William of Ockham: Opera Philosophica et Theologica*. InteLex Corporation.
Park, S. (2011). A confutation of the pessimistic induction. *Journal for General Philosophy of Science*, 42(1):75–84.
Passon, O. (2004). Why isn't every physicist a bohmian? *arXiv preprint* quant-ph/0412119.
Peirce, C. S. (1974). *Collected papers of charles sanders peirce*, volume 5. Harvard University Press.
Popper, K. (2005). *The logic of scientific discovery*. Routledge.
Psillos, S. (2005). Scientific realism and metaphysics. *Ratio*, 18(4):385–404.
Putnam, H. (1975). *Mathematics, Matter and Method: Volume 1, Philosophical Papers*. CUP Archive.
Putnam, H. (1981). *Reason, truth and history*, volume 3. Cambridge University Press.
Pylkkänen, P., Hiley, B., and Pättiniemi, I. (2014). Bohm's approach to quantum mechanics and individuality. *arXiv preprint* arXiv:1405.4772.
Quine, W. V. (1951). Main trends in recent philosophy: Two dogmas of empiricism. *The philosophical review*, pages 20–43.

Roecklein, R. J. (2015). *Locke, Hume, and the Treacherous Logos of Atomism: The Eclipse of Democratic Values in the Early Modern Period*. Lexington Books.

Ross, W. D. (1928). *The works of Aristotle translated into English*. Clarendon Press.

Rovelli, C. (2016). An argument against the realistic interpretation of the wave function. *Foundations of Physics*, 46(10):1229–1237.

Ruark, A. E. (1935). Is the quantum-mechanical description of physical reality complete? *Physical Review*, 48(5):466.

Russell, B. (1940). *An inquiry into meaning and truth*. George Allen and Unwin Ltd.

Sankey, H. (2008). *Scientific realism and the rationality of science*. Ashgate Publishing, Ltd.

Santo, F. D. and Krizek, G. C. (2023). Against the "nightmare of a mechanically determined universe": Why bohm was never a bohmian.

Scheibe, E. (2001). Between rationalism and empiricism. In *Between Rationalism and Empiricism*, pages 1–86. Springer.

Scheibe, E. (2013). *Die Reduktion physikalischer Theorien: Ein Beitrag zur Einheit der Physik*. Springer-Verlag.

Schick, S. (2010). *Contradictio est regula veri: die Grundsätze des Denkens in der formalen, transzendentalen und spekulativen Logik*, volume 53. Meiner Verlag.

Schmidt, H.-J. (2014). Structuralism in physics. In Zalta, E. N., editor, *The Stanford Encyclopedia of Philosophy*. Metaphysics Research Lab, Stanford University, winter 2014 edition.

Schönberger, R. (1990). Realität und differenz. ockhams kritik an der distinctio formalis. In Vossenkuhl, W. and Schönberger, R., editors, *Die Gegenwart Ockhams*, pages 97–122. VCH, Acta Humaniora, Weinheim.

Schröcker, H. (2003). Das verhältnis der allmacht gottes zum kontradiktionsprinzip nach wilhelm von ockham (veröffentlichungen des grabmann-instituts 49).

Sneed, J. D. (1976). Philosophical problems in the empirical science of science: A formal approach. *Erkenntnis*, 10(2):115–146.

Spade, P. V. (1994). *Five texts on the mediaeval problem of universals: Porphyry, Boethius, Abelard, Duns Scotus, Ockham*. Hackett Publishing.

Speziali, P., Einstein, A., and Besso, M. (1972). *Correspondance 1903-1955*. Hermann.

Stanford, K. (2016). Underdetermination of scientific theory. In Zalta, E. N., editor, *The Stanford Encyclopedia of Philosophy*. Metaphysics Research Lab, Stanford University, spring 2016 edition.

Suppe, F. (1977). *The structure of scientific theories*. University of Illinois Press.

Suppe, F. (2000). Understanding scientific theories: An assessment of developments, 1969-1998. *Philosophy of Science*, 67:S102–S115.

Susskind, L. (2016). Copenhagen vs everett, teleportation, and er= epr. *Fortschritte der Physik*, 64(6-7):551–564.

Tarski, A. (1944). The semantic conception of truth: and the foundations of semantics. *Philosophy and phenomenological research*, 4(3):341–376.

Tegmark, M. (2004). Parallel universes. *Science and ultimate reality*, pages 459–491.

Thorburn, W. M. (1918). The myth of occam's razor. *Mind*, pages 345–353.

Timpson, C. G. (2003). The applicability of shannon information in quantum mechanics and zeilinger's foundational principle. *Philosophy of Science*, 70(5):1233–1244.

Timpson, C. G. (2013). *Quantum information theory and the foundations of quantum mechanics*. OUP Oxford.

Tversky, A. and Kahneman, D. (1973). Availability: A heuristic for judging frequency and probability. *Cognitive psychology*, 5(2):207–232.

Vaidman, L. (2016). Many-worlds interpretation of quantum mechanics. In Zalta, E. N., editor, *The Stanford Encyclopedia of Philosophy*. Metaphysics Research Lab, Stanford University, fall 2016 edition.

Valentini, A. (2010). De broglie-bohm pilot-wave theory: Many worlds in denial? *Many worlds*, pages 476–509.

Van Dongen, J. (2010). *Einstein's unification*. Cambridge University Press.

Van Fraassen, B. C. (1980). *The scientific image*. Oxford University Press.

van Inwagen, P. and Sullivan, M. (2017). Metaphysics. In Zalta, E. N., editor, *The Stanford Encyclopedia of Philosophy*. Metaphysics Research Lab, Stanford University, spring 2017 edition.

Visser, M. (1996). Lorentzian wormholes. *From Einstein to Hawking, 412 pp.. AIP Press*, 1.

von Meyenn, K. (2010). *Eine Entdeckung von ganz außerordentlicher Tragweite: Schrödingers Briefwechsel zur Wellenmechanik und zum Katzenparadoxon*. Springer-Verlag.

Von Neumann, J. (2018). *Mathematical foundations of quantum mechanics: New edition*, volume 53. Princeton university press.

von Ockham, W. (1996). *Texte zur Theorie der Erkenntnis und der Wissenschaft*. Reclam.

Wegener, D. T. and Petty, R. E. (1998). The naive scientist revisited: Naive theories and social judgment. *Social Cognition*, 16(1):1–7.

Weihs, G., Jennewein, T., Simon, C., Weinfurter, H., and Zeilinger, A. (1998). Violation of bell's inequality under strict einstein locality conditions. *Physical Review Letters*, 81(23):5039.

Westerhoff, H. V., Winder, C., Messiha, H., Simeonidis, E., Adamczyk, M., Verma, M., Bruggeman, F. J., and Dunn, W. (2009). Systems biology: the elements and principles of life. *FEBS letters*, 583(24):3882–3890.

Whitehead, A. N. and Russell, B. (1912). *Principia mathematica*, volume 2. University Press.

Wigner, E. P. (1960). The unreasonable effectiveness of mathematics in the natural sciences. richard courant lecture in mathematical sciences delivered at new york university, may 11, 1959. *Communications on pure and applied mathematics*, 13(1):1–14.

Wittgenstein, L. (2003). *Tractatus logico-philosophicus, Logisch-philosophische Abhandlung*. Suhrkamp.

Worrall, J. (1989). Structural realism: the best of both worlds? *Dialectica*, 43(1-2):99–124.

Worrall, J. (2007). Miracles and models: Why reports of the death of structural realism may be exaggerated. *Royal Institute of Philosophy Supplement*, 61:125–154.

Zeilinger, A. (1996). *On the interpretation and philosophical foundation of quantum mechanics*. na.

Zeilinger, A. (1999). A foundational principle for quantum mechanics. *Foundations of Physics*, 29(4):631–643.

Zeilinger, A. (2002). Bell's theorem, information and quantum physics. In *Quantum [Un] speakables*, pages 241–254. Springer.

Zeilinger, A. (2003). *Einsteins Schleier: die neue Welt der Quantenphysik*. CH Beck.

SPRINGER NATURE

GPSR Compliance

The European Union's (EU) General Product Safety Regulation (GPSR) is a set of rules that requires consumer products to be safe and our obligations to ensure this.

If you have any concerns about our products, you can contact us on ProductSafety@springernature.com

In case Publisher is established outside the EU, the EU authorized representative is:

Springer Nature Customer Service Center GmbH
Europaplatz 3
69115 Heidelberg, Germany

The manufacturer's authorised representative in the EU is Springer Nature Customer Service Centre GmbH, Europaplatz 3, 69115 Heidelberg, Germany. If you have any concerns regarding our products, please contact ProductSafety@springernature.com

Printed and bound by CPI Group (UK) Ltd, Croydon, CR0 4YY
01/10/2026
02225875-0013